Lecture Notes in Economics and Mathematical Systems

390

Gerhard Sorger

Minimum Impatience Theorems for Recursive Economic Models

Springer-Verlag

Berlin Heidelberg New York
London Paris Tokyo
Hong Kong Barcelona
Budapest

MATH-STAT.

Author

Dr. Gerhard Sorger
Institut für Wirtschaftswissenschaften
Universität Wien
Liechtensteinstr. 13
A-1090 Wien, Austria

This book was published with financial support of the
Österreichische Forschungsgemeinschaft

ISBN 3-540-56022-X Springer-Verlag Berlin Heidelberg New York
ISBN 0-387-56022-X Springer-Verlag New York Berlin Heidelberg

Typesetting: Camera ready by author/editor
42/3140-543210 - Printed on acid-free paper

Editorial

This volume marks an important change in the history of the series **Lecture Notes** in Economics and Mathematical Systems. The managing editors of the series will change: M. Beckmann and W. Krelle, who previously fulfilled this function, will hand over their responsibilities to G. Fandel and W. Trockel.

For more than 25 years, M. Beckmann and W. Krelle have worked as editors for the series, which now numbers almost 400 volumes. M. Beckmann, together with G. Goos and H. P. Künzi, was the founding editor of the series in 1967 and has held the position of managing editor ever since. W. Krelle was a member of the original editorial board, and assumed over the position of managing editor in 1982.

All the managing editors, both old and new, would like to thank all the members of the editorial board for their long standing support of the series, and express the hope that this support will continue in the future. They also wish to thank the authors and editors who have volumes published in the series for submitting their work, and - last but by no means least - the individuals and institutions who buy the books published in the series.

It is the editor's firm conviction that the idea behind the Lecture Notes - to make advanced and specialized topics available quickly, informally, and at a high level to a wide audience - is as important today as it was 25 years ago. The editors hope that the series will continue to be as successful under the new editorship as it has been in the past 25 years.

M. Beckmann   G. Fandel

W. Krelle   W. Trockel

To my father
and the memory of my mother

Preface

This monograph grew out of my effort to better understand the influence of discounting on the qualitative properties of solutions to dynamic optimization problems in economics. Ever since I had read Boldrin and Montrucchio's paper on the Indeterminacy Theorem for the first time I wanted to construct a strictly concave optimal growth model with a "realistic" discount factor and with the most famous chaotic map, the logistic map, as its optimal policy function. Many attempts failed and I suspected that it might perhaps be easier to show that such a model does not exist. So I decided to no longer waste my time on constructing models which would not work and I started to prove my first Minimum Impatience Theorem. After having established this result, a process of continual revisions, generalizations, and improvements set in which finally resulted in the present book.

While I was working on the manuscript I have benefited from encouraging and helpful discussions with A. Araujo, W. Brock, E. Dockner, L. Montrucchio, M. Nermuth, M. Santos, M. Sieveking, and P. Streufert. Most of the manuscript was written while I was working at the Institute for Econometrics, Operations Research, and Systems Theory at the University of Technology in Vienna. I would like to thank all my colleagues there for their kind support. A special note of thanks is due to G. Feichtinger who stimulated my interest in optimal control theory, dynamical systems, and chaos. I would also like to thank A. Novak for helping me to prepare the figures. Of course, all errors in this book are my own responsibility. Financial support from the *Fonds zur Förderung der wissenschaftlichen Forschung* under Grant No. P7783-PHY and from the *Österreichische Forschungsgemeinschaft* under Grant No. 01/0417 is gratefully acknowledged.

Vienna, April 1992

Contents

Chapter 1

Introduction

The purpose of this book is to contribute to the lively discussion about the reasons for and the possible implications of the occurrence of non-linear phenomena and chaos in dynamic economic systems. Since this objective is shared with quite a few recent publications,[1] it is necessary to be more specific and to describe in detail the characteristic features of the present contribution. This will be done in Section 1.1 where we elucidate the basic idea of a Minimum Impatience Theorem by means of a discrete time dynamic programming problem which has been used extensively in the economic growth literature. We shall explain what such a theorem does imply but also what it does not. The four main points made in that section are

 i) that we are interested in *dynamic economic systems*,

 ii) that we discuss the properties of *mathematical models* of the economy and that we do not make any statements about actual time series which have been observed in real economies,

 iii) that we restrict ourselves to *normative* and *deterministic* models as opposed to descriptive or stochastic ones, and

 iv) that our results are more closely related to the *inverse problem* of dynamic optimization than to the dynamic optimization problem itself.

A secondary contribution of the book is that it presents a rather complete theory of deterministic dynamic programming in the framework of a specific model of recursive utility maximization. The model is analysed in both its discrete time and its continuous time version. The necessary and sufficient optimality conditions of dynamic programming are proved from scratch because

---

[1]See [5, 17, 22, 32, 37, 44, 54] for a (certainly not exhaustive) list of books which have the property that one can already tell by their titles that they deal with non-linear dynamics, complexity, and chaos in economics.

these conditions constitute an essential part of the proof of the Minimum Impatience Theorems, and because many of them can be found in the literature only in a form which is either not suitable for our purpose or not valid under our assumptions. Besides, by including every step from the model formulation to the final results the book became more or less self-contained. In Section 1.2 we briefly motivate the choice of recursive utility functionals for our investigation and we present a summary of the contents of the book.

1.1   Minimum Impatience Theorems

Do complicated dynamical patterns occur in economic systems and, if yes, what are those patterns and under what conditions do they emerge? This question was the point of departure for the analysis which finally resulted in the present book. The question, however, can be understood in a number of different ways and it depends on the chosen interpretation which methods have to be applied in order to find an answer. Let us, therefore, briefly outline two main streams of research going into that direction.

The first one considers real economic systems and tries to find out whether the seeming irregularity of observed economic time series is just sheer randomness which can best be captured by stochastic models or whether this irregularity obeys some hitherto unknown but nevertheless simple deterministic laws which are only slightly blurred by random noise. Needless to say that a successful quest for such a law would have tremendous effects on the accuracy of forecasts for economic data. It is also clear, however, that the knowledge of such a law would affect the information structure of the economic system under consideration. It is therefore quite possible that knowing such a law would instantaneously invalidate the very law itself. Without discussing this point any further[2] we just mention that this line of research requires sophisticated econometric methods as well as methods originally developed by physicists (see [17, Chapter 10] or [31] for further details). The Minimum Impatience Theorems developed in the present book will probably have no immediate impact on the progress in this research area.

A second way of answering the aforementioned question is to consider a mathematical model of the economy and to characterize those assumptions which imply or preclude, respectively, the occurrence of dynamically complicated paths in this model. The relevance of any results obtained by this approach depends crucially on the accuracy of the chosen model which is why one should strive for high generality in the model formulation. The methods

---

[2]Current research activities concentrate on finding evidence for the existence of deterministic laws. The problem of characterizing this law is obviously much harder.

employed in this area range from classical results about dynamical systems, differential and difference equations, and bifurcation theory to modern tools from the theory of chaos and strange attractors. There are quite a lot of models for which such an analysis has already been carried out successfully and it is beyond the scope of this introduction to survey the relevant literature.[3] The models can be classified into two different categories: descriptive ones and normative ones. Whereas in a descriptive model the laws which govern the dynamic evolution are assumed to be exogenous, the dynamics of a normative model are endogenously derived from the assumption that economic agents are optimizers. In other words, normative models describe rational behavior in its purest sense. It is this framework of normative models in which the Minimum Impatience Theorems can be applied.

To explain the nature of the Minimum Impatience Theorems in more detail let us consider one particular example of a normative model, an example which has become a standard model in the theory of optimal economic growth. It consists in maximizing the present value of the utility derived from consumption,

$$U(c_1, c_2, c_3, \ldots) = \sum_{t=1}^{\infty} \delta^t u(c_t),$$

subject to the technological constraints

$$\left.\begin{array}{c} x_t = \mu x_{t-1} + f(x_{t-1}) - c_t \\[2mm] x_t \geq 0 \\[2mm] c_t \geq 0 \end{array}\right\} \quad t = 1, 2, 3, \ldots .$$

Here, c_t denotes consumption in period t, x_t denotes the available capital stock at the end of period t, u and f are the short-run utility function and the production function, respectively, and μ and δ are the depreciation factor of capital and the discount factor, respectively. The parameters μ and δ are assumed to be positive and smaller than one. Under appropriate assumptions concerning the continuity, concavity, and boundedness of the functions u and f it can be shown that a unique optimal solution to the above problem exists for every initial capital stock $x_0 \geq 0$. Moreover, it turns out that such an optimal solution can be described by a consumption function $c_t = g(x_{t-1})$ so that the dynamical behavior of the optimal growth paths is completely specified by the recursive equation $x_t = h(x_{t-1})$ with $h(x) = \mu x + f(x) - g(x)$. The function h is called the optimal policy function.

---

[3]The interested reader should consult the bibliographies of the books mentioned in Footnote 1 to get some references on this subject.

The problem of determining the optimal "solution" h from the "givens" u, f, μ, and δ is called an optimal growth problem or, in a more general context, a dynamic programming problem. In order to solve such a problem various sets of necessary optimality conditions have been derived. The purpose of a necessary optimality condition is to deduce useful properties of the unknown solution (in our example this is the optimal policy function h) from the known properties of the problem (that is, from u, f, μ, and δ). Note that this interpretation of a necessary optimality condition is very comprehensive. It includes conditions like the Euler equations or the Bellman equation as well as the Turnpike Theorems. Consider for example the Bellman equation. In many cases it is possible to completely determine h for any given quadruple (u, f, μ, δ) by solving the Bellman equation. In this sense the Bellman equation is complete. The Turnpike Theorems, on the other hand, are not complete since they can only be used to verify one particular property of the solution h: namely, the asymptotic stability of a stationary state.

Now let us view the above problem from a different perspective. Assume that we happen to know a solution h but that we do not know the underlying model, i.e., the quadruple (u, f, μ, δ). How can we derive properties of the model from those of its optimal policy function h? To be more specific, how can we identify or characterize all models (u, f, μ, δ) which satisfy the basic continuity, concavity, and boundedness assumptions and which possess h as an optimal policy function? This problem is usually referred to as the inverse problem of dynamic programming or, when it is restricted to the optimal growth context, as the inverse optimal growth problem. There is by far less literature available on the inverse problem than on the dynamic programming problem itself.[4] However, in a more general class of models than the one defined above (this larger class consists of all so-called reduced form models) one has a remarkable existence theorem for the inverse problem (see [13]). This result, also known as the Indeterminacy Theorem, states that for every function h we can find a reduced form model which admits h as its optimal policy function. Three conditions have to be satisfied in order for the Indeterminacy Theorem to be valid: the function h has to be sufficiently smooth, the capital stocks x_t are required to stay in a compact subset of the Euclidean space, and the discount factor δ must be sufficiently close to zero. At first glance, this result seems to provide an affirmative answer to our question concerning the possibility of chaotic solutions to normative economic models. In fact, one has used the constructive proof of the Indeterminacy Theorem to develop strictly concave deterministic dynamic programming models which are solved by standard examples of chaotic

---

[4][41] and [21] are two useful references on the inverse optimal growth problem. Whereas the former is restricted to deterministic models, the latter deals with a more general stochastic framework. Both papers consider only one-sector growth models.

dynamical systems like the logistic map in discrete time (see [13]) or the Lorenz equations in continuous time (see [48]). The third assumption of the Indeterminacy Theorem, however, implies that these chaotic solutions are optimal only at rather extreme and unrealistic values for the discount factor δ. There are two possible explanations for this drawback: either the special construction used in the proof of the Indeterminacy Theorem is not powerful enough to produce models with realistic parameter values or the very nature of the chaotic time paths precludes their optimality in any such model however cleverly it might be constructed. This is the point where the Minimum Impatience Theorems come into play.

Whereas the Indeterminacy Theorem provides sufficient conditions for the existence of a solution to the inverse problem of dynamic programming, the Minimum Impatience Theorems can be regarded as necessary existence conditions for this problem. One can use them to obtain an upper bound on the discount factor in every dynamic programming problem which generates a given solution h as its optimal policy function. More specifically, our results provide a means to find for every mapping h a number δ^h such that h is certainly not an optimal policy function in any model (u, f, μ, δ) satisfying the continuity, concavity, and boundedness assumptions mentioned earlier as well as the inequality $\delta \geq \delta^h$. The bound δ^h can be equal to one, in which case we cannot conclude anything useful, but it will usually be strictly smaller than one provided that the dynamics generated by h are sufficiently complicated and non-monotonic. A Minimum Impatience Theorem is not a complete necessary condition, that is, we cannot derive from it the exact model (u, f, μ, δ) generating h as its optimal solution but only an estimate on the discount factor δ. The incompleteness is not surprising because one can easily see that the inverse dynamic programming problem does not have a unique solution.

Several different Minimum Impatience Theorems will be proved in the course of this book. In Chapter 4 we consider models which are formulated in discrete time and in Chapter 7 we present results for the continuous time case. We derive results which are based on global properties of the given optimal solution h as well as results which get by with the knowledge of local properties. We prove Minimum Impatience Theorems which focus on the non-monotonicity of optimal growth paths and others which are derived from the sensitive dependence of these paths with respect to initial conditions. All our results will be illustrated by simple examples and they are valid for a class of dynamic optimization models which is even more general than the class of reduced form optimal growth models considered in [13] or [48].

At this stage, a more detailed discussion of the Minimum Impatience Theorems does not seem to be expedient. We shall continue it in Section 3.3 when the basic notation has been introduced and the model formulation has been completed. Let us just add one more remark concerning the term "Minimum

Impatience". It is derived from the well known fact that the discount factor δ is a measure of the impatience of the decision maker. The smaller the discount factor, the higher is her/his rate of impatience. In the limiting case where δ is equal to zero we have an infinitely impatient agent who completely disregards the future. Such an agent is said to be myopic. Now suppose that we observe the outcome of a decision process in a black box. In terms of the optimal growth model discussed earlier the observed outcome is a path $x_0, x_1, x_2, \ldots$ of capital stocks and the black box contains the unknown production technology (f, μ), the unknown utility function u, and the unknown discount factor δ. We would like to know if it is possible to recognize the decision maker's time-preference rate or, equivalently, the discount factor δ by the observed outcome. Although the Minimum Impatience Theorems cannot provide a full answer to this question for all possible growth paths $x_0, x_1, x_2, \ldots$, they give us conclusive evidence for a high time-preference rate in many cases in which the outcome exhibits a complicated dynamical pattern. Because of this reason it would have been justified to call our results "Theorems of Revealed Time-Preference". However, we have chosen the denotation "Minimum Impatience Theorems" in order to indicate that only a lower bound on the time-preference rate can be obtained. At this point we would like to emphasize once again that this lower bound depends neither on the particular production function f nor on the utility function u.

From the above discussion one might get the impression that complicated dynamical behavior can be optimal in the class of normative economic models under consideration only for unrealistic parameter values. This is not correct. Let us, therefore, conclude this section by pointing out what the Minimum Impatience Theorems do *not* imply. Analogously to the necessary optimality conditions of dynamic programming which proceed on the assumption of a given model (u, f, μ, δ), the Minimum Impatience Theorems as necessary existence conditions for the inverse dynamic programming problem proceed on the assumption of a given solution h. Therefore, we can make a statement about possible parameter values only for this particular function h. We cannot conclude that every chaotic map requires a low discount factor to become an optimal policy function. Neither can we make this claim for less complicated dynamical solution patterns like limit cycles. In particular, our results do not contradict the provable optimality of limit cycles at all possible discount factors.[5] Let us also emphasize once more that the bound δ^h mentioned earlier

---

[5] See [9] for the continuous time case and [12, p. 638] for the discrete time case. In these papers it is shown that, for every discount factors $\delta \in (0, 1)$, one can construct a model (u, f, μ, δ) which has a stable limit cycle as an optimal growth path. They do not show, however, that the same limit cycle, i.e., the same policy function h, is optimal for all $\delta \in (0, 1)$. Only this stronger property could be a contradiction to the Minimum Impatience Theorems.

can be equal to one in some cases so that we cannot conclude anything useful. This can happen, for example, if the trajectories generated by h are monotonic and/or stable. However, if the dynamics of h are sufficiently complex then a useful bound $\delta^h < 1$ will be obtained. We think that the examples discussed in this book, for which the Minimum Impatience Theorems are conclusive (i.e., $\delta^h < 1$), provide enough justification to present our approach. In this regard it should also be noted that even the mere fact that such a bound δ^h exists for some policy functions h has not been known before. Our results provide therefore a new theoretical insight into the qualitative structure of solutions to discounted deterministic dynamic programming problems.

1.2   Recursive Economic Models

Consider once more the simple optimal growth model formulated in the previous section. The objective functional U of this model satisfies

$$U(c_1, c_2, c_3, \ldots) = u(c_1) + \delta U(c_2, c_3, c_4, \ldots).$$

Functionals with this property are said to be additively separable and they have long been criticized as they imply that the decision maker has a constant rate of time-preference. A more general class of utility functionals, known as recursive utility functionals, was introduced by Koopmans in [39]. In this class the aforementioned property is replaced by the weaker statement that

$$U(c_1, c_2, c_3, \ldots) = A(c_1, U(c_2, c_3, c_4, \ldots))$$

where A is called the aggregator function. Thus, we still have a weak form of separability between current consumption and total future utility but it need no longer be additive and linear. Moreover, Koopmans showed that every such functional necessarily exhibits "impatience" (in a sense defined precisely in [39]) over some regions of its domain. The fact that both the theory of recursive utility functionals and the Minimum Impatience Theorems are centered around the notion of impatience calls for a combination of the results from these two areas. Therefore, we have chosen to develop our results in the general framework of dynamic optimization problems which are formulated in terms of recursive utility functionals. The first step towards this goal was a rigorous derivation of the dynamic programming techniques for such models which are essential for the proof of the Minimum Impatience Theorems. The presentation of these techniques is also a contribution of this book, although a less important one.

As everyone can tell from the table of contents, the book consists of two parts which are almost identical in structure. The first one deals with models in a discrete time framework whereas the second part assumes that time is a continuous variable. Except for the technical details both parts proceed along the same lines which is why we treat the continuous time case in considerable less detail than the discrete time case. In particular, we provide more intuition and motivation, a greater number of examples, and more detailed explanations of the proofs in Part 1 than we do in Part 2. At many places in Part 2 we simply omit entire proofs which are similar or identical to their discrete time counterparts.

Each part starts with a chapter entitled "Model Formulation" in which we describe the dynamic constraints as well as the preference structure of the models. Unlike most of the work on recursive preferences and their application in optimal growth theory our models are stated in the so-called reduced form. By this we mean that the model does not contain any control variables (like consumption) explicitly but that it is formulated entirely in terms of state variables (capital stocks). From a technical point of view we would like to mention that our utility functionals are what has been called β-myopic by Boyd [16]. This property implies a certain degree of impatience along all feasible state trajectories. Stating it in another way, the maximal discount factor along an arbitrary trajectory[6] cannot exceed a given constant, namely, β^{-1}. Whereas β-myopic utility functionals have been considered previously in the discrete time framework (see [16]), their treatment in continuous time models seems to be new.

The second chapter of each part is devoted to the preparation of "Preliminary Results" which will be needed later to prove the Minimum Impatience Theorems. First, we discuss the Bellman equation of dynamic programming and use it to derive necessary and sufficient optimality conditions. Whereas in the continuous time setting we have to restrict ourselves to concave models, no such restriction is necessary in discrete time models. Another minor blemish of the results for the continuous time case is that the optimality conditions are valid only under an additional interiority assumption. These differences between discrete time and continuous time models are mainly caused by the well known fact that optimality conditions for continuous time models require stronger smoothness hypotheses than the corresponding conditions for discrete time models. Among the "Preliminary Results" there are also two transformations of dynamic programming models. The first one, called a translation, leaves the set of optimal solutions invariant but changes some monotonicity properties of the utility function and the optimal value function. The second transformation acts like a quick-motion apparatus by transforming a model with the optimal policy function h into one that has a higher iterate of h as its optimal policy function. This transformation is called the time-τ transformation. In

---

[6]Note that the discount factor in a model with a recursive utility functional is not a constant but varies with time and with the chosen growth path.

Part 1 there is also a brief section on "Indeterminacy and Turnpikes" which continues the discussion and the motivation of the Minimum Impatience Theorems already commenced in Section 1.1. Moreover, we present examples of strictly concave optimal growth models which are solved by the logistic map $h(x) = 4x(1-x)$ and the tent map $h(x) = 1 - |2x - 1|$, respectively. The discount factors used in these models are much higher than those of similar examples previously reported in the literature.

The main results of the book are contained in Chapter 4 for the discrete time case and in Chapter 7 for continuous time models. In each of these chapters we present two central results from which various other Minimum Impatience Theorems are then derived as corollaries. The first of these results shows that highly non-monotonic optimal growth paths require a high rate of impatience whereas the second one relates the sensitive dependence of optimal growth paths with respect to initial conditions to the discount rate. Examples are included to illustrate the application of the results.

In the discrete time case we devote a whole section to one-dimensional models. This has two reasons: on the one hand, our results are very easy to apply in the one-dimensional case and, on the other hand, the class of one-dimensional maps contains the best known examples of chaotic dynamical systems like the aforementioned logistic map and the tent map.

It will be seen that a certain set-valued mapping, the so-called average correspondence, is of central importance for a whole class of Minimum Impatience Theorems. Its properties are discussed in a separate section before we proceed to derive the local versions of the Minimum Impatience Theorems. These theorems can be used to obtain upper bounds on the minimal discount factor without utilizing global properties of the given policy function h. More specifically, in order to apply a local Minimum Impatience Theorem all we need to know is the Jacobian matrix of h at one of its fixpoints, say, $\bar{x}$. If this matrix has the property of "expansiveness" or if one of its eigenvalues is real with multiplicity one and absolute value greater than one, then a useful bound $\delta^h < 1$ can be obtained. The notion of expansiveness combines a strong form of local instability of the fixpoint $\bar{x}$ with the local non-monotonicity of the trajectories of h starting close to $\bar{x}$. In most cases it is possible to generalize the local Minimum Impatience Theorems to periodic points of arbitrary period τ. The main tool for this generalization is the aforementioned time-τ transformation. It is used to construct an auxiliary problem which is defined on the same state space as the original one and which has a fixpoint instead of the τ-periodic point as one of its solutions. Curiously, the auxiliary problem is a discrete time model no matter how the original model was formulated (in discrete time or in continuous time).

Section 4.5 (no corresponding section is included in Part 2) concludes the presentation of the Minimum Impatience Theorems and outlines some perhaps

10

fruitful directions for further research. We also discuss ways in which our results would have to be generalized in order to qualify as true necessary optimality conditions for chaos. Essentially, such a generalization would entail the inclusion of a defining property of chaos into the calculation of the upper bound δ^h described earlier. The existence of positive Ljapunov exponents for the mapping h is probably the most promising candidate for such a property.

Part I

Discrete Time Models

Chapter 2

Model Formulation

In this chapter, we define the class of recursive economic models which provides the framework for the analysis in the first part of the book. Typical examples of such models occur quite naturally in the theory of optimal economic growth. Therefore, we have chosen to adopt the whole terminology from economic growth theory and, in particular, to call the problem under consideration an optimal growth model. Some standard assumptions which are characteristic of optimal growth theory, however, like the non-negativity of capital stocks or the hypothesis of free disposal will not be needed to derive any of the results. Consequently, we omit these assumptions from the model formulation or we replace them by more general ones. This enables the application of the methods and results developed in later chapters also to situations which are outside the realm of optimal economic growth theory.

We begin with a discussion of the constraints which are typically encountered in dynamic economic models and which, in the optimal growth context, represent the available production technology. These constraints define the set of feasible solutions **F** which we call the set of feasible growth paths. Some topological properties of **F** are presented in Section 2.1.

In order to have a well defined optimization problem we must also specify a criterion functional. This is done in Section 2.2 where we use the aggregator approach originally developed by Lucas and Stokey [45] to define recursive utility functionals in the sense of Koopmans [39]. The assumptions imposed on the aggregator functions are similar to those used by Boyd [16] and they imply the existence of so-called β-myopic utility functionals. Unlike the bulk of the hitherto existing literature on recursive preferences we define the utility functionals not on the set of possible consumption streams but directly on the set of feasible growth paths, **F**. In addition to being more general than the usual approach, this provides a natural link between the technological growth conditions and the time-perspective implicitly defined by the utility functionals.

Section 2.3 integrates the technological constraints and the preferences into the complete model and introduces the optimal value function. Since consumption does not explicitly occur in our model, we obtain what is called a reduced form model of optimal economic growth. It is shown that, for every feasible initial capital stock, there exists an optimal solution in this model. Finally, we discuss the notion of the minimal discount factor along a given feasible growth path, a concept which will become of vital importance in the Minimum Impatience Theorems.

Throughout this part of the book we consider optimal growth models in a discrete time framework. Consequently, the time variable t will always denote a number in the set of non-negative integers $\mathbf{Z}_0^+ = \{0, 1, 2, \ldots\}$. Moreover, we denote by $\mathbf{Z}^+ = \{1, 2, 3, \ldots\}$ the set of positive integers. For every $n \in \mathbf{Z}^+$, the symbol $\mathbf{R}^n$ denotes the n-dimensional Euclidean space. The scalar product of two vectors x and y in $\mathbf{R}^n$ is denoted by $x \cdot y$. If $x \in \mathbf{R}^n$, then $|x|$ denotes the Euclidean norm of x, i.e., $|x| = (x \cdot x)^{1/2}$. For sequences $(x_0, x_1, x_2, \ldots)$ and $(x_t, x_{t+1}, x_{t+2}, \ldots)$ we shall frequently use the short hand forms $_0x$ and $_tx$, respectively. Given these notations, the interpretation of expressions like $_0x = (x_0, {}_1x)$ should be obvious.

2.1　Technology

The feasible solutions in an optimal growth problem are the possible growth paths, i.e., those sequences of capital stocks which can occur under the given technological constraints of the production processes. The feasible set of solutions is therefore determined by the production technology. We assume that there are n different capital goods, where $n \in \mathbf{Z}^+$ is arbitrary but given. Consequently, we can describe the state of the economy at any time $t \in \mathbf{Z}_0^+$ by a state vector $x_t \in \mathbf{R}^n$ representing the available stocks of the capital goods. The interpretation of the state vectors as capital stocks would require that x_t is an element of the positive orthant in $\mathbf{R}^n$. This restriction to non-negative state vectors, however, will not be needed for our purpose and we replace it by the more general concept of a state space. We denote this state space by X and assume that it is a non-empty, closed, and convex subset of $\mathbf{R}^n$. Elements of X are called feasible state vectors or feasible vectors of capital stocks.

The technology is assumed to be stationary. By this we mean that the set of possible capital stocks at time $t + 1$ depends only on the state of the economy in the previous period t. This assumption implies that the technological constraints of the economy can be described by a production technology set $T \subseteq X \times X$. Formally, we require that the pair (x_t, x_{t+1}) is an element of T for every period $t \in \mathbf{Z}_0^+$. We assume that the set T is closed and convex and that it satisfies a growth condition in the sense that there exist real numbers $\alpha \geq 0$ and $\beta \geq 0$ such that $|x'| \leq \alpha + \beta|x|$ holds for all $(x, x') \in T$. This assumption is most likely satisfied in any real problem. Note also that feasible growth paths need not be bounded if $\beta \geq 1$. Finally, it will be assumed that for every feasible capital stock $x \in X$ there is at least one possible capital stock x' for the next period, i.e., the x-section of T defined by $T_x = \{x' \in X \mid (x, x') \in T\}$ is non-empty for every $x \in X$.

We summarize the notations and assumptions introduced so far in the following definition.

Definition 2.1 Let $n \in \mathbf{Z}^+$ be given. A set $X \subseteq \mathbf{R}^n$ is called a *state space* if it is non-empty, closed, and convex. A set $T \subseteq X \times X$ is said to represent a *technology* on the state space X if the following conditions are satisfied:

i) The set T is closed and convex.

ii) The set $T_x = \{x' \in X \mid (x, x') \in T\}$ is non-empty for every $x \in X$.

iii) There exist real numbers $\alpha \geq 0$ and $\beta \geq 0$ such that the inequality $|x'| \leq \alpha + \beta |x|$ holds for all $(x, x') \in T$.

The number β referred to in part iii) of the above definition represents an upper bound on the growth factor of any feasible growth path in the technology T (see Lemma 2.2 below). Therefore, we call T a *β-bounded* technology. Note also that the growth condition iii) implies that the set $T \cap (C \times X)$ is bounded whenever C is a bounded subset of X. Some basic properties of a technology are presented in the following lemma.

Lemma 2.1 *If T is a technology on the state space $X \subseteq \mathbf{R}^n$, then the x-section T_x is a non-empty, compact, and convex set for every $x \in X$. Moreover, it holds that the set-valued mapping $x \mapsto T_x$ is continuous on X.*

PROOF. Non-emptiness, compactness, and convexity of T_x are immediate consequences of Definition 2.1. The continuity of the correspondence T_x follows from Theorems 3.4 and 3.5 in [63]. □

We can now define rigorously what we mean by a feasible growth path.

Definition 2.2 A *feasible growth path* for a given technology T on the state space $X \subseteq \mathbf{R}^n$ is any sequence $_0x = (x_0, x_1, x_2, \ldots)$ of vectors $x_t \in X$ such that $(x_t, x_{t+1}) \in T$ holds for all $t \in \mathbf{Z}_0^+$. The set of feasible growth paths will be denoted by $\mathbf{F}$. Moreover, for every feasible state vector $x \in X$ we denote by $\mathbf{F}(x)$ the subset of $\mathbf{F}$ consisting of those feasible growth paths $_0x \in \mathbf{F}$ which satisfy the initial condition $x_0 = x$.

In what follows it will be necessary to endow the set of feasible growth paths **F** with a topological structure. To this end, we define for every positive real number β the β-norm of a sequence $_0x = (x_0, x_1, x_2, \ldots) \in (\mathbf{R}^n)^\infty$ by[1]

$$\|_0x\|_\beta = \sup \left\{ |x_t|/\beta^t \,\middle|\, t \in \mathbf{Z}_0^+ \right\}.$$

The sequence space $\mathbf{S}_\beta$ is defined as

$$\mathbf{S}_\beta = \left\{ _0x \in (\mathbf{R}^n)^\infty \,\middle|\, \|_0x\|_\beta < \infty \right\}.$$

It is easy to see that $(\mathbf{S}_\beta, \|\cdot\|_\beta)$ is a Banach space. The norm topology on $\mathbf{S}_\beta$ is also called the β-*topology*. Moreover, it is obvious that for every $\beta' \geq \beta$ we have $\|_0x\|_{\beta'} \leq \|_0x\|_\beta$. This has two important consequences: first, it shows that $\mathbf{S}_\beta \subseteq \mathbf{S}_{\beta'}$ and, second, it shows that convergence in the β-topology implies convergence in the β'-topology, i.e., the β-topology on $\mathbf{S}_\beta$ is stronger than the relative β'-topology on $\mathbf{S}_\beta$. Note also that the β-topology is stronger than the relative product topology since convergence in the β-norm implies pointwise convergence. Regarding the set of feasible growth paths **F** we have the following result.

Lemma 2.2 *Let* **F** *be the set of feasible growth paths in a β-bounded technology* T *on the state space* $X \subseteq \mathbf{R}^n$ *and let* $x \in X$ *be an arbitrary feasible state vector. Then it holds that* **F** *and* **F**(x) *are non-empty and convex sets. If $\beta > 1$, then* **F** *and* **F**(x) *are subsets of* $\mathbf{S}_\beta$ *which are closed in the β-topology.*

PROOF. The fact that **F** and **F**(x) are non-empty sets for all $x \in X$ follows from the definition of a feasible growth path and from part ii) of Definition 2.1. Convexity of **F** and **F**(x) follows immediately from convexity of X and T. Now assume that $\beta > 1$. If $_0x = (x_0, x_1, x_2, \ldots) \in \mathbf{F}$ then it must hold that $|x_{t+1}| \leq \alpha + \beta|x_t|$ for all $t \in \mathbf{Z}_0^+$. It is easy to see that this implies

$$|x_t| \leq \alpha \sum_{s=0}^{t-1} \beta^s + \beta^t |x_0| = \beta^t \left(\frac{\alpha}{\beta - 1} + |x_0| \right) - \frac{\alpha}{\beta - 1}$$

for all $t \in \mathbf{Z}_0^+$. This shows that the β-norm of $_0x$ satisfies

$$\|_0x\|_\beta \leq \frac{\alpha}{\beta - 1} + |x_0| < \infty. \tag{2.1}$$

In particular it follows that $_0x \in \mathbf{S}_\beta$ which proves that **F** is a subset of $\mathbf{S}_\beta$.

---

[1]The β-norms should not be confused with the more familiar p-norms defined by $\|_0x\|_p = \left[\sum_{t=0}^\infty |x_t|^p \right]^{1/p}$ for all $p \in [1, \infty)$.

Finally, let $_0x^{(k)} \in \mathbf{F}$ be a feasible growth path for every $k \in \mathbf{Z}^+$ and assume that $_0x^{(k)}$ converges in the β-topology to an element $_0\bar{x} \in \mathbf{S}_\beta$ as k approaches infinity. Because the β-topology is stronger than the topology of pointwise convergence, it follows that $\lim_{k\to\infty} x_t^{(k)} = \bar{x}_t$ for all $t \in \mathbf{Z}_0^+$. Since X and T are closed sets, we can conclude that $\bar{x}_t \in X$ and $(\bar{x}_t, \bar{x}_{t+1}) \in T$ for all $t \in \mathbf{Z}_0^+$. This shows that $_0\bar{x} \in \mathbf{F}$ which implies that $\mathbf{F}$ is closed in the β-topology. The closedness of the sets $\mathbf{F}(x)$ for all $x \in X$ is proved analogously. $\qquad\square$

In the case where $\beta > 1$ it follows from Lemma 2.2 that $\mathbf{F}$ is a complete metric space when it is endowed with the relative β-topology. In the following we shall always assume that this is the case unless explicitly stated otherwise. For example, continuity of functionals on $\mathbf{F}$ will be understood with respect to the relative β-topology.

We conclude this section by a result which will be useful for proving the existence of optimal growth paths.

Lemma 2.3 *Let $\mathbf{F}$ be the set of feasible growth paths in a β-bounded technology T on the state space $X \subseteq \mathbf{R}^n$. If $1 < \beta < \beta'$ and $x \in X$, then it holds that $\mathbf{F}(x)$ is a subset of $\mathbf{S}_{\beta'}$ which is compact in the β'-topology.*

PROOF. The inclusion $\mathbf{F}(x) \subseteq \mathbf{S}_{\beta'}$ follows from Lemma 2.2 and from $\mathbf{S}_\beta \subseteq \mathbf{S}_{\beta'}$. Moreover, it is shown in [16, Lemma 2] that the β'-topology and the product topology coincide on $\mathbf{F}(x)$. Using $|x_t| \leq \beta^t \|_0x\|_\beta$ and (2.1) it follows from Tychonoff's theorem that $\mathbf{F}(x)$ is compact in the product topology. This completes the proof. $\qquad\square$

2.2 Preferences

Having defined the set of feasible growth paths, $\mathbf{F}$, we now proceed to specify a criterion functional U on this set which assigns a unique utility level to each path $_0x \in \mathbf{F}$. A standard method of doing this is to specify U by

$$U(_0x) = \sum_{t=0}^{\infty} \delta^t V(x_t, x_{t+1}) \qquad (2.2)$$

where $V : T \mapsto \mathbf{R}$ denotes the short-run utility function and $\delta \in (0,1)$ the discount factor. Functionals defined in this way are time-additive and exhibit a constant rate of time-preference $\rho = (1/\delta) - 1 > 0$. Although these properties are very convenient from a mathematical point of view, they represent a rather unrealistic picture of the actual behavior of economic agents. More general utility functionals, commonly referred to as recursive utility functionals, were introduced by Koopmans and his collaborators in [39, 40]. Lucas and Stokey [45] considered bounded recursive utility functionals and applied a fixpoint argument to show that they can be determined by their so-called aggregator functions. The aggregator approach has only recently been generalized to a larger class of (possibly unbounded) utility functionals by Boyd [16]. The material in the present section consists to some extent of modifications of the results in [16]. There is, however, one important difference between our formulation and the one commonly used in the literature: namely, that we define the utility functionals on the set of feasible growth paths and not on the set of feasible consumption paths. In this sense our model can be called a reduced form model of optimal economic growth with recursive preferences, that is, a model in which consumption has already been "maximized out".[2] One advantage of the reduced form approach is that it provides a direct link between the characteristics of the technology and those of the utility functionals in a natural way. Another approach which relates the definition of recursive utility to the underlying production technology can be found in Streufert's work [64, 65].

Before we present the details let us briefly explain the idea. Our goal is to define a functional $U : \mathbf{F} \mapsto \mathbf{R}$ which has two important properties. The first one is continuity with respect to the relative β-topology on $\mathbf{F}$. This property was termed β-myopia in [16] and it reflects the fact that total utility is not too sensitive with respect to changes of the capital stocks in the distant future.[3] To put it differently, β-myopia presupposes a certain degree of impatience on the side of the decision maker. The second property we shall require from the utility functional is that it is *recursively defined*. This means that the utility of a given feasible growth path depends only on the (investment and consumption) decisions made in the present period and on the total utility derived from future consumption. One of the most important advantages of formulating an optimal growth problem with a recursive utility functional is that the model becomes amenable to the methods of dynamic programming as will be shown in Section 3.1.

We have already mentioned that we require the utility functional U to be a β-continuous functional from $\mathbf{F}$ to the real line. Let us denote by $\mathcal{C}(\mathbf{F})$ the set of all such functionals. Assume that $\beta > 1$ and define the functional $\phi \in \mathcal{C}(\mathbf{F})$ by

$$\phi(_0x) = 1 + \|_0x\|_\beta.$$

Since $\mathbf{F} \subseteq \mathbf{S}_\beta$, it follows that the functional ϕ is well defined on $\mathbf{F}$. Moreover, we have the following property.

---

[2]A reduced form model with recursive utility functionals very similar to ours has been used in a recent paper by Dana and Le Van [26].

[3]Strictly speaking, this is only true if $\beta > 1$, a condition which will be assumed throughout the remainder of Part 1.

Lemma 2.4 *Let* $\mathbf{F}$ *be the set of feasible growth paths in a β-bounded technology* T *with* $\beta > 1$. *For every* $_0x = (x_0, {_1}x) \in \mathbf{F}$ *it holds that* $\phi({_1}x)/\phi({_0}x) < \beta$.

PROOF. We have

$$\frac{\phi({_1}x)}{\phi({_0}x)} = \frac{1 + \sup_{t \geq 1} |x_t|/\beta^{t-1}}{1 + \sup_{t \geq 0} |x_t|/\beta^t} \leq \frac{1 + \sup_{t \geq 1} |x_t|/\beta^{t-1}}{1 + \sup_{t \geq 1} |x_t|/\beta^t} = \frac{1 + z\beta}{1 + z}$$

where $z = \sup_{t \geq 1} |x_t|/\beta^t \geq 0$. Because of $\beta > 1$, the last expression in the above chain of inequalities is smaller than β which proves the lemma. $\square$

A functional $f \in \mathcal{C}(\mathbf{F})$ will be called ϕ-*bounded*, if

$$\sup \left\{ |f({_0}x)|/\phi({_0}x) \,\Big|\, {_0}x \in \mathbf{F} \right\} < \infty.$$

Consider the subspace $\mathcal{C}_\phi(\mathbf{F}) \subseteq \mathcal{C}(\mathbf{F})$ consisting of all ϕ-bounded continuous functionals on $\mathbf{F}$ and define a norm on $\mathcal{C}_\phi(\mathbf{F})$ by

$$\|f\| = \sup \left\{ |f({_0}x)|/\phi({_0}x) \,\Big|\, {_0}x \in \mathbf{F} \right\}$$

for all $f \in \mathcal{C}_\phi(\mathbf{F})$. It is easy to show that $(\mathcal{C}_\phi(\mathbf{F}), \|\cdot\|)$ is isometrically isomorph to the Banach space of bounded continuous functionals on $\mathbf{F}$ and, hence, it is itself a Banach space. In particular, $\mathcal{C}_\phi(\mathbf{F})$ is a complete metric space, a property which will be used in the proof of Theorem 2.2 below. Furthermore, we shall need the following contraction mapping theorem which is proved in [16].[4]

Theorem 2.1 *Let* $T : \mathcal{C}_\phi(\mathbf{F}) \mapsto \mathcal{C}(\mathbf{F})$ *be an operator such that the following conditions are true:*

i) *There exists a functional* $\bar{f} \in \mathcal{C}_\phi(\mathbf{F})$ *such that* $T[\bar{f}] \in \mathcal{C}_\phi(\mathbf{F})$.

ii) $T[f] \leq T[f']$ *whenever* f *and* f' *are elements of* $\mathcal{C}_\phi(\mathbf{F})$ *satisfying* $f \leq f'$.

iii) *There exists a positive real number* $\theta < 1$ *such that for all real numbers* $z > 0$ *and for all* $f \in \mathcal{C}_\phi(\mathbf{F})$ *it holds that* $T[f + z\phi] \leq T[f] + z\theta\phi$.

Then it follows that $T[\mathcal{C}_\phi(\mathbf{F})] \subseteq \mathcal{C}_\phi(\mathbf{F})$ *and that* T *is a contraction mapping on* $\mathcal{C}_\phi(\mathbf{F})$ *with contraction factor* θ.

---

[4]The reader might note that condition i) in Theorem 2.1 is slightly weaker than the corresponding assumption in the Weighted Contraction Mapping Theorem in [16]. Except for an obvious modification, however, the proof of Theorem 2.1 is the same as the one in [16] and will not be repeated here.

Using the above theorem we can now prove the existence of a recursive utility functional $U \in \mathcal{C}(\mathbf{F})$ which is generated by a continuous function $A : T \times \mathbf{R} \mapsto \mathbf{R}$, called the aggregator function. The idea is to define U recursively by[5]

$$U(_0x) = A((x_0, x_1), U(_1x)) \tag{2.3}$$

for all $_0x \in \mathbf{F}$. This is equivalent to the condition that U is a fixpoint of the operator $\mathcal{T}_A : \mathcal{C}_\phi(\mathbf{F}) \mapsto \mathcal{C}(\mathbf{F})$ defined by

$$\mathcal{T}_A[f](_0x) = A((x_0, x_1), f(_1x)). \tag{2.4}$$

If we can verify that the operator $\mathcal{T}_A$ satisfies the conditions of Theorem 2.1, then the existence of a unique solution to the equation $\mathcal{T}_A[f] = f$ is guaranteed by Banach's fixpoint theorem. In the following definition we specify assumptions on the aggregator function which will then be shown to imply the required properties of the operator $\mathcal{T}_A$. Except for the necessary modifications to adapt them to the reduced form model, these assumptions are identical to a special case of the conditions stated in [16] for the non-reduced form model. The reader should note that the conditions are cardinal in nature so that the corresponding utility function has to be understood as a cardinal one.

Definition 2.3 Let T be a given β-bounded technology on the state space $X \subseteq \mathbf{R}^n$ and assume $\beta > 1$. A continuous function $A : T \times \mathbf{R} \mapsto \mathbf{R}$ is called an *aggregator function* for T if the following conditions are satisfied:

i) There exists a vector $\bar{v} \in \mathbf{R}^n$ such that the mapping $_0x \mapsto A((x_0, x_1), \bar{v} \cdot x_1)$ from $\mathbf{F}$ to $\mathbf{R}$ is ϕ-bounded.

ii) The function $z \mapsto A((x, x'), z)$ from $\mathbf{R}$ to $\mathbf{R}$ is non-decreasing for all $(x, x') \in T$ and there exists a real number $\bar{\delta} \in (0, 1)$ such that the Lipschitz condition

$$|A((x, x'), z) - A((x, x'), z')| \leq \bar{\delta}|z - z'| \tag{2.5}$$

holds for all $(x, x') \in T$ and for all $z, z' \in \mathbf{R}$.

iii) The parameters β and $\bar{\delta}$ satisfy the relation $\beta\bar{\delta} < 1$.

Comparing Definition 2.3 with the corresponding definition in Boyd [16] it seems that the boundedness condition i) is more general. Given the Lipschitz condition (2.5), however, one could assume without loss of generality that $\bar{v} = 0$ which is exactly Boyd's case. We have chosen not to do so because the additional flexibility provided by the unspecified vector $\bar{v}$ will be useful in Section 3.2. There is also another formulation of this condition which is easier to verify since it does not involve the space $\mathbf{F}$. It is presented in the following lemma.

---

[5]Note that $_0x = (x_0, _1x) = (x_0, x_1, x_2, \ldots)$.

Lemma 2.5 *Condition i) of Definition 2.3 is satisfied if and only if there exist a vector $\bar{v} \in \mathbf{R}^n$ and a constant $M' > 0$ such that the inequality*

$$|A((x, x'), \bar{v} \cdot x')| \leq M'(1 + |x|)$$

holds for all $(x, x') \in T$.

PROOF. The "if" part is obvious from the definition of ϕ-boundedness. To prove the "only if" part we assume the contrary, i.e., that there exist two sequences of state vectors $x^{(0)}, x^{(1)}, x^{(2)}, \ldots$ and $y^{(0)}, y^{(1)}, y^{(2)}, \ldots$ such that for all $m \in \mathbf{Z}^+$ it holds that $(x^{(m)}, y^{(m)}) \in T$ and

$$|A((x^{(m)}, y^{(m)}), \bar{v} \cdot y^{(m)})| > m(1 + |x^{(m)}|).$$

From Lemma 2.2 it follows that we can find feasible growth paths $_0x^{(m)} \in \mathbf{F}(x^{(m)})$ with $x_0^{(m)} = x^{(m)}$ and $x_1^{(m)} = y^{(m)}$. Since we know that the functional $_0x \mapsto A((x_0, x_1), \bar{v} \cdot x_1)$ is ϕ-bounded, there must exist a constant $M' > 0$ such that

$$M'(1 + \|_0x^{(m)}\|_\beta) \geq |A((x^{(m)}, y^{(m)}), \bar{v} \cdot y^{(m)})| > m(1 + |x^{(m)}|)$$

for all $m \in \mathbf{Z}^+$. Using (2.1) we obtain

$$M'\left(1 + \frac{\alpha}{\beta - 1} + |x^{(m)}|\right) > m(1 + |x^{(m)}|)$$

for all $m \in \mathbf{Z}^+$. It is easy to see that this is a contradiction and, hence, it follows that the lemma is proven. $\qquad\square$

The next theorem is the main result of this section.

Theorem 2.2 *Let T be a β-bounded technology on the state space $X \subseteq \mathbf{R}^n$ and assume that $\beta > 1$. Moreover, let $A : T \times \mathbf{R} \mapsto \mathbf{R}$ be an aggregator function for this technology. Then there exists a unique utility functional $U \in \mathcal{C}_\phi(\mathbf{F})$ such that Equation (2.3) is satisfied. Moreover, it holds that*

$$U(_0x) = \lim_{k \to \infty} \mathcal{T}_A^{(k)}[f](_0x) \tag{2.6}$$

for all $_0x \in \mathbf{F}$ and for all $f \in \mathcal{C}_\phi(\mathbf{F})$. Here, $\mathcal{T}_A^{(k)} : \mathcal{C}_\phi(\mathbf{F}) \mapsto \mathcal{C}(\mathbf{F})$ denotes the k-th iterate of the operator $\mathcal{T}_A$ defined in (2.4).

PROOF. We verify that the operator $\mathcal{T}_A$ satisfies the conditions of Theorem 2.1. Defining the functional $\bar{f} \in \mathcal{C}_\phi(\mathbf{F})$ by $\bar{f}(_0x) = \bar{v} \cdot x_1$, one can easily see that the assumption $\mathcal{T}_A[\bar{f}] \in \mathcal{C}_\phi(\mathbf{F})$ is equivalent to condition i) of Definition 2.3. The monotonicity assumption ii) of Theorem 2.1 follows from the monotonicity of the function $z \mapsto A((x, x'), z)$. Finally, we have for all $z > 0$

$$
\begin{aligned}
\mathcal{T}_A[f + z\phi]({}_0x) &= A((x_0, x_1), f({}_1x) + z\phi({}_1x)) \\
&\leq A((x_0, x_1), f({}_1x)) + \bar{\delta}z\phi({}_1x) \\
&= \mathcal{T}_A[f]({}_0x) + z\bar{\delta}\frac{\phi({}_1x)}{\phi({}_0x)}\phi({}_0x).
\end{aligned}
$$

Because of condition iii) of Definition 2.3 and because of Lemma 2.4 it follows that assumption iii) of Theorem 2.1 is satisfied with $\theta = \beta\bar{\delta} < 1$. Hence, we conclude that $\mathcal{T}_A$ is a contraction operator on $\mathcal{C}_\phi(\mathbf{F})$. Since $\mathcal{C}_\phi(\mathbf{F})$ is a complete metric space, Banach's fixpoint theorem shows that there exists a unique fixpoint $U \in \mathcal{C}_\phi(\mathbf{F})$ which can be calculated by the method of successive approximations, i.e., Equation (2.6) holds for all ${}_0x \in \mathbf{F}$ and for all $f \in \mathcal{C}_\phi(\mathbf{F})$. This completes the proof. $\square$

The following result provides a useful estimate on the utility functional U described in Theorem 2.2.

Corollary 2.1 *If the assumptions of Theorem 2.2 are satisfied, then there exists a constant $M > 0$ such that, for every $x \in X$ and for every feasible path ${}_0x \in \mathbf{F}(x)$, the inequality $|U({}_0x)| \leq M(1 + |x|)$ holds.*

PROOF. Since U is ϕ-bounded on $\mathbf{F}$, there exists $M' > 0$ such that the inequality $|U({}_0x)| \leq M'(1 + \|{}_0x\|_\beta)$ holds for all ${}_0x \in \mathbf{F}(x)$. Combining this with the estimate given in (2.1) proves the result. $\square$

Theorem 2.2 states that the utility functional U is β-myopic. The following lemma improves this result by showing that U is in fact β'-myopic for all $\beta' < 1/\bar{\delta}$.

Lemma 2.6 *Let the assumptions of Theorem 2.2 be satisfied and assume $\beta' \in (1, \bar{\delta}^{-1})$. Then it follows that the utility functional U is continuous in the β'-topology on $\mathbf{F}$.*

PROOF. Let ${}_0x \in \mathbf{F}$ be an arbitrary feasible growth path. The lemma is proven if we can show that for every $\epsilon > 0$ one can find a real number $\gamma > 0$ such that the following is true: for all ${}_0y \in \mathbf{F}$ satisfying $\|{}_0x - {}_0y\|_{\beta'} < \gamma$ it holds that $|U({}_0x) - U({}_0y)| < \epsilon$. To this end, we first note that for all ${}_0y \in \mathbf{F}$ with $\|{}_0x - {}_0y\|_{\beta'} < 2$ we have $|x_t - y_t| < 2(\beta')^t$ for all $t \in \mathbf{Z}_0^+$. From this fact and from Corollary 2.1 we can conclude that there exists a constant $M > 0$ such that the following inequality holds for all $t \in \mathbf{Z}_0^+$:

$$
\begin{aligned}
\bar{\delta}^t|U({}_tx) - U({}_ty)| &\leq \bar{\delta}^t M(2 + |x_t| + |y_t|) \\
&\leq 2\bar{\delta}^t M \left[1 + |x_t| + (\beta')^t\right] \\
&\leq 2\bar{\delta}^t M \left[1 + \beta^t\|{}_0x\|_\beta + (\beta')^t\right]
\end{aligned}
$$

Because of $\max\{\beta, \beta'\} < 1/\bar{\delta}$ and $\bar{\delta} < 1$ this shows that there exists a constant $s \in \mathbf{Z}^+$ (depending only on the path $_0x$ but not on γ as long as $\gamma \in (0, 2)$) such that

$$\bar{\delta}^t |U(_tx) - U(_ty)| \le \frac{\epsilon}{2} \tag{2.7}$$

for all $t \ge s$. Continuity of the aggregator function A implies that we can choose the number $\gamma \in (0, 2)$ so small that

$$|A((x_{t-1}, x_t), U(_tx)) - A((y_{t-1}, y_t), U(_tx))| < \frac{\epsilon}{2} \frac{1 - \bar{\delta}}{1 - \bar{\delta}^s} \tag{2.8}$$

whenever $t \le s$ and $\|_0x - _0y\|_{\beta'} < \gamma$. Finally, we can use Equation (2.3) and the Lipschitz condition (2.5) to obtain

$$
\begin{aligned}
|U(_0x) &- U(_0y)| \\
&= |A((x_0, x_1), U(_1x)) - A((y_0, y_1), U(_1y))| \\
&\le \bar{\delta}|U(_1x) - U(_1y)| + |A((x_0, x_1), U(_1x)) - A((y_0, y_1), U(_1x))|.
\end{aligned}
$$

Proceeding in this way it follows that

$$
\begin{aligned}
|U(_0x) - U(_0y)| &\le \bar{\delta}^s |U(_sx) - U(_sy)| \\
&+ \sum_{t=1}^{s} \bar{\delta}^{t-1} |A((x_{t-1}, x_t), U(_tx)) - A((y_{t-1}, y_t), U(_tx))|.
\end{aligned}
$$

Substituting (2.7) and (2.8) into this inequality we obtain $|U(_0x) - U(_0y)| < \epsilon$ which completes the proof. □

The value of the aggregator function $A((x, x'), z)$ measures the maximal total utility that can be derived from consumption under the given technological constraints if the current capital stock is x, if the capital stock in the next period is required to be x', and if the total utility from the next period onwards is given by z.

A particularly important case arises if the function A is additively separable, i.e., if there exists a function $V : T \mapsto \mathbf{R}$ and a constant $\delta \in [0, 1)$ such that

$$A((x, x'), z) = V(x, x') + \delta z \tag{2.9}$$

for all $(x, x') \in T$ and for all $z \in \mathbf{R}$. The corresponding time-additive utility functional is given by Equation (2.2) and has been used extensively in the economic literature (see, e.g., [63]).

So far, we have not made any assumptions on the aggregator function A which guarantee that the utility functional U associated with A is concave. This will be done in the following lemma.

Lemma 2.7 *Let the assumptions of Theorem 2.2 be satisfied and assume that the aggregator function $A : T \times \mathbf{R} \mapsto \mathbf{R}$ is concave. Then it holds that $U : \mathbf{F} \mapsto \mathbf{R}$ is a concave functional.*

PROOF. It is easy to verify that the concavity of A and the monotonicity of the aggregator function as specified in assumption ii) of Definition 2.3 imply that the operator $\mathcal{T}_A$ preserves concavity. Therefore, the concavity of U follows from the approximation result (2.6) if one starts from an arbitrary concave functional $f \in \mathcal{C}_\phi(\mathbf{F})$, for example, from $f \equiv 0$. □

Concavity of the aggregator function A is sufficient but not necessary for the concavity of the utility functional U. A simple example for a non-concave aggregator function which generates a concave utility functional is the additively separable aggregator

$$A((x, x'), z) = -a|x|^2 + b|x'|^2 + \delta z$$

where the parameters a, b, and δ are all positive and satisfy the restriction $a\delta \geq b$. The corresponding utility functional,

$$U(_0x) = -a|x_0|^2 - \sum_{t=1}^{\infty} \delta^t(a - b/\delta)|x_t|^2,$$

is obviously concave although the aggregator function A is not. Another example is the Uzawa aggregator

$$A((x, x'), z) = (z - 1)e^{-V(x,x')}$$

where $V : T \mapsto \mathbf{R}$ is non-negative and concave (see [66]). The utility functional generated by this aggregator is

$$U(_0x) = -\sum_{t=0}^{\infty} \exp\left[-\sum_{s=0}^{t} V(x_s, x_{s+1})\right].$$

We conclude this section by pointing out that the conditions summarized in Definition 2.3 are not the weakest possible ones to guarantee the existence of a recursive utility functional. In fact, Boyd [16] imposes more general assumptions on the aggregator function which could have been adapted to the reduced form model under consideration. Also Streufert's conditions used in [65] are more general. We have chosen this particular set of assumptions because the notational and technical cost of the added generality seemed to be too high. Moreover, the assumptions of Definition 2.3 appeared to be natural, given that the set of feasible growth paths $\mathbf{F}$ is known to be a subset of $\mathbf{S}_\beta$. In this regard it should be noted that in our model a uniform summability condition is already "built in" since for every feasible growth path $_0x \in \mathbf{F}$ it holds that the infinite series

$$\sum_{t=0}^{\infty} \bar{\delta}^t x_t$$

is absolutely convergent. This property follows simply from Lemma 2.2 and from condition iii) of Definition 2.3. Analogously, one can see that

$$\sum_{t=0}^{\infty} \bar{\delta}^t f(x_t)$$

converges absolutely for all functions $f : X \mapsto \mathbf{R}$ which satisfy the growth condition

$$\sup \left\{ |f(x)|/(1 + |x|) \,\Big|\, x \in X \right\} < \infty.$$

We shall make use of these properties in later chapters.

2.3   Optimal Growth Paths and Impatience

In the preceding two sections we have described a model of the technological constraints of an economy and we have explained how one can define intertemporal preferences on the corresponding set of feasible growth paths. This is all that is needed to completely specify an optimal growth model.

Definition 2.4 Let $X \subseteq \mathbf{R}^n$ be a non-empty, closed, and convex set. An *optimal growth model* with state space X is a pair (T, A) where T is a β-bounded technology on X with $\beta > 1$ and $A : T \times \mathbf{R} \mapsto \mathbf{R}$ is an aggregator function for this technology. The optimal growth model is said to have *additively separable* preferences if there exists a constant $\delta > 0$ and a function $V : T \mapsto \mathbf{R}$ such that Equation (2.9) holds. The optimal growth model is said to be *concave* if the aggregator function A is concave.

Wherever in Part 1 of this book we talk about an optimal growth model it should be understood in the sense of this definition. Of course, this does not mean that there are no other mathematical models which can adequately describe the process of optimal capital accumulation. As a matter of fact, there are quite a lot of such models and many of them do not belong to the class described in Definition 2.4. The reason why we introduced this terminology is merely to have a concise notation for a model which satisfies the assumptions on the technology and the preferences discussed in Sections 2.1 and 2.2, respectively. In this regard we would like to emphasize once more that the results presented in this book are not confined to optimal growth theory because the assumptions imposed on an "optimal growth model" (T, A) are general enough to capture a lot of other situations as well.

Now recall from Definition 2.2 that $\mathbf{F}(x)$ is the set of those feasible growth paths $_0x \in \mathbf{F}$ which satisfy the initial condition $x_0 = x$. Here, $x \in X$ is an arbitrary vector of feasible capital stocks available at the beginning of the first period $(t = 0)$. The function W defined on the set of feasible capital stocks X by

$$W(x) = \sup \left\{ U(_0x) \,\Big|\, _0x \in \mathbf{F}(x) \right\} \tag{2.10}$$

is called the *optimal value function* of the problem (T, A). The following result shows that this function is well defined and finite on X, and that it is a concave function whenever the model (T, A) is concave.

Lemma 2.8 *Let (T, A) be an optimal growth model with state space $X \subseteq \mathbf{R}^n$ and utility functional $U : \mathbf{F} \mapsto \mathbf{R}$. The optimal value function W is well defined by Equation (2.10) and everywhere finite, that is, $W : X \mapsto \mathbf{R}$. More specifically, it holds that $|W(x)| \le M(1 + |x|)$ for all $x \in X$, where $M > 0$ is a constant independent of x. If the model (T, A) is concave, then it follows that the optimal value function W is also concave.*

PROOF. From Lemma 2.2 it follows that $\mathbf{F}(x)$ is non-empty for all $x \in X$. Together with Corollary 2.1 this implies that the supremum in Equation (2.10) is finite and that the optimal value function satisfies the inequality $|W(x)| \le M(1 + |x|)$. This proves the first part of the lemma. The second assertion follows from well known results on general concave maximization problems and from Lemma 2.7. □

Having defined the optimal value function, it is clear what we mean by an optimal growth path. Furthermore, it follows from our previous results that optimal growth paths do exist for all initial states $x \in X$. These issues are dealt with in the following definition and in Lemma 2.9.

Definition 2.5 Let (T, A) be an optimal growth model with state space X, utility functional U, and optimal value function W. Moreover, let $x \in X$ be a feasible initial state. An *optimal growth path* for the initial state x is a feasible growth path $_0x \in \mathbf{F}(x)$ such that $U(_0x) = W(x)$.

Lemma 2.9 *Let (T, A) be an optimal growth model with state space X. Then there exists an optimal growth path for every initial state $x \in X$.*

PROOF. In view of condition iii) of Definition 2.3 we can find a real number β' satisfying $\beta < \beta' < 1/\bar{\delta}$. From Lemma 2.3 it follows that $\mathbf{F}(x)$ is a compact set in the β'-topology, and from Lemma 2.6 we know that the utility function U is continuous with respect to the β'-topology on $\mathbf{F}$. Therefore, the existence of an optimal growth path is a consequence of Weierstrass' existence theorem. □

It should be noted that in view of the above lemma we can replace the supremum in Equation (2.10) by a maximum. This will help to simplify some of the proofs in later chapters.

The Minimum Impatience Theorems which will be proved in Chapter 4 are necessary optimality conditions which rule out the occurrence of certain dynamically complicated optimal growth paths in the entire class of optimal growth problems under consideration. One cannot expect such a result to hold if this class consists of all concave optimal growth models. For example, consider the additively separable aggregator function A defined in (2.9) with discount factor $\delta > 0$ and assume that the short-run utility function V is constant on T. This implies that the corresponding utility functional U from (2.2) and the optimal value function W are also constant (on $\mathbf{F}$ and on X, respectively). Furthermore, it is obvious that (T, A) is a concave optimal growth model. On the other hand, because the utility functional is constant, every feasible growth path maximizes U and, consequently, we cannot prevent complicated growth paths from becoming optimal. This type of indeterminacy can be ruled out by requiring that the short-run utility function V satisfies a slightly stronger property than concavity (which, however, is still not quite as strong as strict concavity; see Definition 2.6 below). Given this assumption, one can easily see that the utility functional U is strictly concave and that optimal growth paths are uniquely determined by their initial state x_0.

As a second example consider again an aggregator function A which is additively separable and assume that the discount factor δ in (2.9) is equal to zero. In this case, only the first period decision matters since no weight at all is given to future utility. Consequently, all feasible growth paths for which the first period capital stocks coincide give rise to the same utility no matter how complicated the dynamical patterns in the remaining periods are. As the preceding example, this one is also characterized by a great deal of indeterminacy. This situation as well as similar ones can be excluded by the assumption that the mapping $z \mapsto A((x, x'), z)$ is strictly increasing instead of only non-decreasing as it was required in Definition 2.3.

In view of the two examples just discussed we shall concentrate our analysis in Chapter 4 to the class of strictly concave optimal growth models as specified in the following definition.

Definition 2.6 An optimal growth model (T, A) on the state space $X \subseteq \mathbf{R}^n$ is called *strictly concave* if the following two conditions are satisfied:

i) The aggregator function $A : T \times \mathbf{R} \mapsto \mathbf{R}$ is a concave function and the strict inequality

$$A\Big(((1 - \lambda)x + \lambda\bar{x}, (1 - \lambda)x' + \lambda\bar{x}'), (1 - \lambda)z + \lambda\bar{z}\Big)$$
$$> (1 - \lambda)A((x, x'), z) + \lambda A((\bar{x}, \bar{x}'), \bar{z})$$

holds for all $z \in \mathbf{R}$ and $\bar{z} \in \mathbf{R}$, for all $\lambda \in (0, 1)$, and for all pairs $(x, x') \in T$ and $(\bar{x}, \bar{x}') \in T$ which satisfy $x \neq \bar{x}$.

ii) The function $z \mapsto A((x, x'), z)$ from $\mathbf{R}$ to $\mathbf{R}$ is strictly increasing for all $(x, x') \in T$.

It is obvious that a model (T, A) is strictly concave if its aggregator function A is strictly concave. The converse, however, is not necessarily true since the condition stated in part i) of the above definition is weaker than strict concavity of A. An important implication of an optimal growth model being strictly concave is that its optimal value function W is strictly concave. This will be proved in Lemma 3.4. It can also be shown that the utility function U of a strictly concave model is a strictly concave functional on the set of feasible paths $\mathbf{F}$. Let us emphasize once more, however, that strict concavity of (T, A) is not necessary for strict concavity of W and U.

Now let us turn to a slightly different issue. Assume that the agent who controls the economy has decided on a specific growth path $_0x = (x_0, x_1, x_2, \ldots)$ so that her/his total utility is given by $U(_0x) = A((x_0, x_1), z)$ with $z = U(_1x)$. Assume further that someone offers the agent a certain amount $\zeta > 0$ of utility for the next period.[6] How does this affect the agent's total utility? If she/he does not modify the production plans, then the total utility will be changed to $U'(_0x) = A((x_0, x_1), z + \zeta)$ which is greater than or equal to $U(_0x)$ because of the monotonicity condition ii) of Definition 2.3. On the other hand, the Lipschitz condition (2.5) for the aggregator function implies that $U'(_0x) \leq U(_0x) + \bar{\delta}\zeta$ which shows that the prospective additional utility will be discounted by a factor $\delta \leq \bar{\delta}$. More specifically, we have

$$\delta = \delta(_0x, \zeta) = \frac{U'(_0x) - U(_0x)}{\zeta} = \frac{A((x_0, x_1), z + \zeta) - A((x_0, x_1), z)}{\zeta}$$

which shows that the discount factor δ depends on the growth path $_0x$ as well as on the amount ζ offered. It is quite likely that the relative value of having the amount ζ now as compared to having it in the next period increases with ζ. This would imply that the function $\delta(_0x, \zeta)$ is decreasing with respect to ζ.

---

[6]Since we are dealing with a cardinal utility function it makes sense to talk about amounts of utility.

Moreover, it shows that the actual degree of impatience of the agent along the path $_0x$ can be measured by the infimum of the discount factors $\delta(_0x, \zeta)$ over all admissible amounts $\zeta \in \mathbf{R}$. This motivates the following definition of the minimal discount factor as an indicator of the agent's degree of impatience.

Definition 2.7 Let (T, A) be an optimal growth model on the state space $X \subseteq \mathbf{R}^n$ and let $U : \mathbf{F} \mapsto \mathbf{R}$ be its utility function. Moreover, let $D \subseteq \mathbf{F}$ be a given non-empty set of feasible paths for (T, A). Define $\underline{\delta}(D)$ as the infimum over all $_0x \in D$, $t \in \mathbf{Z}_0^+$, and $\zeta \in \mathbf{R} \backslash \{0\}$ of the expression

$$\frac{A((x_t, x_{t+1}), U(_{t+1}x) + \zeta) - A((x_t, x_{t+1}), U(_{t+1}x))}{\zeta}.$$

We call $\underline{\delta}(D)$ the *minimal discount factor* of the aggregator function A on the set D. If D coincides with the entire feasible set $\mathbf{F}$, then we say that $\underline{\delta}(\mathbf{F})$ is the *globally minimal discount factor* of A.

Several remarks are in order concerning this definition. First of all, it is obvious from the monotonicity properties of an aggregator function that $0 \leq \underline{\delta}(D) \leq \bar{\delta}$ holds for all subsets D of $\mathbf{F}$. Furthermore, for an additively separable aggregator function (see Equation (2.9)) we have $\underline{\delta}(D) = \bar{\delta} = \delta$ which shows that $\underline{\delta}(D)$ is independent of its argument D. In the general case, however, $\underline{\delta}(D)$ depends on D. The additional flexibility obtained by the introduction of the set D in Definition 2.7 will lead to sharper results in Chapter 4. In fact, most of our results will make statements about the minimal discount factor along a single feasible path $_0x = (x_0, x_1, x_2, \ldots)$, that is, we shall specify the set D by $D = \{_0x\}$.

We would also like to mention that it is common to define the discount rate ρ corresponding to the discount factor δ by $\rho = (1/\delta) - 1$. It is therefore sensible to call $\bar{\rho}(D) = [1/\underline{\delta}(D)] - 1$ the maximal discount rate or the maximal rate of impatience of the aggregator function A on the set D. However, in the discrete time framework it is easier to work with the discount factor than with the discount rate which is why we shall state all our results in terms of the minimal discount factor $\underline{\delta}(D)$.

Chapter 3

Preliminary Results

This chapter serves two different purposes. The first one is to present a number of results which will be used in the proofs of the Minimum Impatience Theorems and the second one is to motivate the Minimum Impatience Theorems by discussing two important results from the literature pertaining to the qualitative structure of optimal growth paths.

The methods of dynamic programming for the reduced form optimal growth model with recursive preferences are presented in Section 3.1. First we show that the optimal value function is a solution of the functional equation of dynamic programming, i.e., of the Bellman equation. In addition, we show that the optimal value function is continuous and that it is the only solution of the Bellman equation in the class of all functions which satisfy a linear growth condition. Then, a standard argument is used to prove that the optimal growth paths of a strictly concave optimal growth model are the trajectories of a dynamical system. The link between the analysis of optimal growth paths and the theory of dynamical systems is the optimal policy function which relates to every feasible state vector, x_t, the optimal capital stock for the next period, x_{t+1}. The central results of Section 3.1, however, are the necessary and sufficient optimality conditions which we derive from the Bellman equation. No differentiability or concavity assumptions are required for these results. The special formulation of the optimality conditions, which might look unfamiliar at first glance, is chosen with regard to their later use in the proofs of the Minimum Impatience Theorems.

Section 3.2 begins with a discussion of some results on equivalent optimal growth models, that is, on models which have the same set of optimal growth paths. It is shown how one can construct an equivalent model by a simple transformation of the aggregator function which we call a "translation". Translations are shown to preserve several basic properties of the optimal growth model. Among these properties are strict concavity and additive separability. Moreover, we prove that for every concave optimal growth model there exists an equivalent model such that the optimal value function of the new model attains its maximum at a predetermined point in the state space. Another useful transformation for optimal growth models is the time-τ transformation. Applied to a model (T, A) with optimal policy function h it generates a new model (T', A') with optimal policy function $h^{(\tau)}$, the τ-th iterate of h. The time-τ transformation does also preserve strict concavity and additive separability.

In Section 3.3 we present two results from the optimal growth literature which help to motivate the specific question we are trying to answer by the Minimum Impatience Theorems. These results are the Indeterminacy Theorem and the Turnpike Theorem. Both theorems make a statement about the possible dynamical behavior of optimal paths in strictly concave optimal growth models. The Indeterminacy Theorem states that virtually every behavior however complicated it might be can be optimal in a strictly concave optimal growth model provided that the feasible state space X is compact and that the rate of impatience of the decision maker is sufficiently high or, equivalently, that the minimal discount factor of the aggregator function is sufficiently close to zero. The Turnpike Theorem, on the other hand, shows that for a given strictly concave optimal growth model with additively separable preferences[1] the dynamical behavior of an optimal growth path is "simple" provided that the discount factor is sufficiently close to one. The Minimum Impatience Theorems are motivated as a converse to the Indeterminacy Theorem on the one hand and as a complement to the Turnpike Theorem on the other hand.

3.1 Dynamic Programming

In this section we present necessary and sufficient optimality conditions for the optimal growth problem (T, A). These conditions are based on Lemma 3.1 which is the Bellman equation of dynamic programming. Before we formulate it let us introduce some notation. Define the function $\psi : X \mapsto \mathbf{R}$ by

$$\psi(x) = 1 + |x|$$

for all $x \in X$. We say that a function $f : X \mapsto \mathbf{R}$ is ψ-bounded if[2]

$$\sup \left\{ |f(x)| / \psi(x) \,\middle|\, x \in X \right\} < \infty.$$

The set of all ψ-bounded functions on X will be denoted by $\mathcal{B}_\psi(X)$. It follows from Lemma 2.8 that the optimal value function of an optimal growth model is ψ-bounded. We can now prove the Bellman equation.

---

[1]By this we mean that the short-run utility function V in Equation (2.9) is given but not the discount factor δ.

[2]The reader should note the similarity to the notion of ϕ-boundedness introduced in Section 2.2.

Lemma 3.1 *Let (T, A) be an optimal growth model with state space $X \subseteq \mathbf{R}^n$ and denote by W the optimal value function of this model. Then it holds that W is the unique function in the space $\mathcal{B}_\psi(X)$ which satisfies the following functional equation for all $x \in X$:*

$$W(x) = \sup \left\{ A((x, x'), W(x')) \,\Big|\, x' \in T_x \right\} \tag{3.1}$$

PROOF. Let $x \in X$ be an arbitrary state vector. According to Lemma 2.9 there exists a growth path $_0x \in \mathbf{F}(x)$ such that $U(_0x) = W(x)$. Moreover, it is obvious that the inequality $U(_1x) \leq W(x_1)$ must hold where $_0x = (x, {_1x})$ and $_1x = (x_1, x_2, x_3, \ldots)$. Using these properties, Equation (2.3), and the montonicity of the aggregator function, we obtain

$$\begin{aligned}
W(x) &= U(_0x) \\
&= A((x, x_1), U(_1x)) \\
&\leq A((x, x_1), W(x_1)) \\
&\leq \sup \left\{ A((x, x'), W(x')) \,\Big|\, x' \in T_x \right\}.
\end{aligned} \tag{3.2}$$

On the other hand, we also know that for every fixed $x' \in T_x$ there exists a path $_1x' = (x', x_2', x_3', \ldots) \in \mathbf{F}(x')$ with $U(_1x') = W(x')$. In view of the definition of the optimal value function and Equation (2.3) this implies for the growth path $_0x' = (x, {_1x'}) \in \mathbf{F}(x)$ that

$$W(x) \geq U(_0x') = A((x, x'), U(_1x')) = A((x, x'), W(x')).$$

It follows immediately from this inequality and from (3.2) that the optimal value function W is a solution to (3.1).

Now assume that $V \in \mathcal{B}_\psi(X)$ is another solution to (3.1). Choose $x_0 \in X$ and $\epsilon > 0$ arbitrarily. Then it must hold for all $x' \in T_{x_0}$ that $V(x_0) \geq A((x_0, x'), V(x'))$. On the other hand, from (3.1) it follows that there exists a vector $x' \in T_{x_0}$ such that $W(x_0) \leq A((x_0, x'), W(x')) + \epsilon$. Taking both of these inequalities together and using (2.5) we obtain

$$W(x_0) - V(x_0) \leq A((x_0, x'), W(x')) - A((x_0, x'), V(x')) + \epsilon \leq \bar{\delta}|W(x') - V(x')| + \epsilon.$$

By interchanging the roles of W and V one can also show that there exists $x'' \in T_{x_0}$ such that

$$V(x_0) - W(x_0) \leq \bar{\delta}|W(x'') - V(x'')| + \epsilon.$$

Defining $x_1 \in T_{x_0}$ either by $x_1 = x'$ or by $x_1 = x''$, whichever yields a larger value of $|W(x_1) - V(x_1)|$, we obtain

$$|W(x_0) - V(x_0)| \leq \bar{\delta}|W(x_1) - V(x_1)| + \epsilon.$$

Proceeding in this way we can construct a feasible path $_0x = (x_0, x_1, x_2, \ldots) \in \mathbf{F}(x_0)$ such that

$$|W(x_t) - V(x_t)| \leq \bar{\delta}|W(x_{t+1}) - V(x_{t+1})| + \epsilon$$

holds for all $t \in \mathbf{Z}_0^+$. It is easy to see that this implies that the inequality

$$|W(x_0) - V(x_0)| \leq \bar{\delta}^t|W(x_t) - V(x_t)| + \epsilon \sum_{s=0}^{t-1} \bar{\delta}^s \leq \bar{\delta}^t|W(x_t) - V(x_t)| + \frac{\epsilon}{1 - \bar{\delta}}$$

holds for all $t \in \mathbf{Z}_0^+$. Since both W and V are assumed to be ψ-bounded, we conclude from this inequality that there exists a constant $M > 0$ which is independent of t such that

$$|W(x_0) - V(x_0)| \leq \bar{\delta}^t M(1 + |x_t|) + \frac{\epsilon}{1 - \bar{\delta}}$$

holds for all $t \in \mathbf{Z}_0^+$. From (2.1) it follows that there exists another constant $M' > 0$ such that $|x_t| \leq M'(1 + |x_0|)\beta^t$ for all $t \in \mathbf{Z}_0^+$. Substituting this into the above inequality yields

$$|W(x_0) - V(x_0)| \leq \bar{\delta}^t M \left[1 + M'(1 + |x_0|)\beta^t\right] + \frac{\epsilon}{1 - \bar{\delta}}$$

for all $t \in \mathbf{Z}_0^+$. Because of $\bar{\delta} < 1$ and $\beta\bar{\delta} < 1$ (see Definition 2.3) we obtain for $t \to \infty$ that

$$|W(x_0) - V(x_0)| \leq \frac{\epsilon}{1 - \bar{\delta}}.$$

Since $\epsilon > 0$ and $x_0 \in X$ were chosen arbitrarily, it follows that $W(x) = V(x)$ for all $x \in X$. Therefore, W is the only function in $\mathcal{B}_\psi(X)$ which satisfies Equation (3.1). $\square$

An equivalent formulation of Lemma 3.1 is that the optimal value function is the unique fixpoint of the operator T_A^* defined on the set of all ψ-bounded functions, $\mathcal{B}_\psi(X)$, by

$$T_A^*[f](x) = \sup \left\{ A((x, x'), f(x')) \,\Big|\, x' \in T_x \right\} \tag{3.3}$$

for all $f \in \mathcal{B}_\psi(X)$ and for all $x \in X$. We can use this fact to prove that the optimal value function of (T, A) is continuous.

Lemma 3.2 *The optimal value function W of an optimal growth model (T, A) with state space $X \subseteq \mathbf{R}^n$ is continuous on X.*

PROOF. Consider the space $\mathcal{C}_\psi(X) \subseteq \mathcal{B}_\psi(X)$ of all real-valued, ψ-bounded, and continuous functions on X and define a norm $\|\cdot\|_\psi$ on $\mathcal{C}_\psi(X)$ by

$$\|f\|_\psi = \sup \left\{ |f(x)|/\psi(x) \,\Big|\, x \in X \right\}.$$

The space $(\mathcal{C}_\psi(X), \|\cdot\|_\psi)$ is isometrically isomorph to the Banach space of bounded and continuous functions on X and, hence, it is itself a Banach space. In a way similar to the proof of the Weighted Contraction Mapping Theorem in [16] one can show that $\mathcal{T}_A^*[f]$ is ψ-bounded for all $f \in \mathcal{C}_\psi(X)$. Moreover, it follows from the Maximum Theorem of Berge [10] (see also [63, Theorem 3.6]) and from the continuity and compactness of T_x (see Lemma 2.1) that $\mathcal{T}_A^*[f]$ is continuous whenever f is continuous. Therefore, we have $\mathcal{T}_A^*[\mathcal{C}_\psi(X)] \subseteq \mathcal{C}_\psi(X)$. Without presenting the details[3] we just mention that it is possible to show that $\mathcal{T}_A^*$ is a contraction mapping. Consequently, we can conclude that its fixpoint W is unique and that it must be an element of $\mathcal{C}_\psi(X)$. This completes the proof of the lemma. □

We note that there is another way to prove the above lemma. In fact, one could apply the Maximum Theorem directly to the optimal growth problem (T, A) rather than to the problem of maximizing $A((x, x'), f(x'))$ over the set T_x. This approach is feasible since we know from Lemma 2.3 that the correspondence $x \mapsto \mathbf{F}(x)$ is compact-valued with regard to the β'-topology for some $\beta' \in (\beta, \bar{\delta}^{-1})$ and from Lemma 2.6 that U is continuous with respect to that topology. The only thing which remains to be shown is that the correspondence $x \mapsto \mathbf{F}(x)$ is also β'-continuous.

It should be emphasized that the validity of the Bellman equation has to be distinguished from the validity of Bellman's optimality principle.[4] However, if the aggregator function is strictly monotonic with respect to future utility, then the optimality principle defined below is valid.

Definition 3.1 Let (T, A) be an optimal growth model with state space $X \subseteq \mathbf{R}^n$ and feasible set $\mathbf{F}$. We say that the *optimality principle* holds in this model if, for all optimal paths $_0x = (x_0, x_1, x_2, \ldots) \in \mathbf{F}$ and for all $t \in \mathbf{Z}^+$, the feasible path $_tx = (x_t, x_{t+1}, x_{t+2}, \ldots)$ is also optimal.

Lemma 3.3 *Let (T, A) be an optimal growth model with state space $X \subseteq \mathbf{R}^n$ and assume that the function $z \mapsto A((x, x'), z)$ is strictly increasing on $\mathbf{R}$ for all $(x, x') \in T$. Then the optimality principle holds in (T, A).*

---

[3] See [63, Theorem 4.6] for a similar argument in a slightly different framework.

[4] This can be seen from the following example: $n = 1$, $X = [0, 1/2]$, $T = X \times X$, $A((x, x'), z) = xz + x'$. The path $_0x = (0, 1/2, 0, 0, 0, \ldots)$ is optimal for the initial state 0 but the path $_1x = (1/2, 0, 0, 0, \ldots)$, which is an infinite tail of $_0x$, is not optimal for its initial state $1/2$. Therefore, the optimality principle in the sense of Definition 3.1 does not hold.

PROOF. The proof is by contradiction. Assume that there exists a number $t \in \mathbf{Z}^+$ and an optimal growth path $_0x = (x_0, x_1, x_2, \ldots) \in \mathbf{F}$ such that the path $_tx$ is not optimal. Without loss of generality we may assume that $t = 1$. Since $_1x$ is not optimal, there exists another feasible growth path $_1x' = (x_1, x_2', x_3', \ldots) \in \mathbf{F}(x_1)$ with initial capital stock x_1 such that $U(_1x') > U(_1x)$. Using the strict monotonicity assumption on the aggregator function and Equation (2.3) one can see that this implies $U(_0x') > U(_0x)$ where $_0x' = (x_0, _1x')$. This is a contradiction to the optimality of $_0x$ and the lemma is proven. $\qquad \square$

An important special case in which the optimality principle holds is the one of a strictly concave optimal growth model (see Definition 2.6). This case exhibits also some other properties which deserve special attention. We deal with two of these properties in the subsequent two lemmas. First we prove that the optimal value function of a strictly concave optimal growth model is strictly concave.

Lemma 3.4 *Let (T, A) be a strictly concave optimal growth model on the state space $X \subseteq \mathbf{R}^n$. Then it follows that the optimal value function W is strictly concave on X.*

PROOF. From Lemma 3.1 we know that Equation (3.1) holds for all $x \in X$. Moreover, the supremum in this equation is attained by some vector $x' \in T_x$ because A and W are continuous functions and because T_x is a compact set. Let x and $\bar{x}$ be two different state vectors and let $\lambda \in (0, 1)$ be given. Then we can find $x' \in T_x$ and $\bar{x}' \in T_{\bar{x}}$ such that $W(x) = A((x, x'), W(x'))$ and $W(\bar{x}) = A((\bar{x}, \bar{x}'), W(\bar{x}'))$. Because of the concavity of W we have

$$W((1 - \lambda)x' + \lambda\bar{x}') \geq (1 - \lambda)W(x') + \lambda W(\bar{x}').$$

Using these properties and the one stated in part i) of Definition 2.6 we obtain

$$
\begin{aligned}
W((1 &- \lambda)x + \lambda\bar{x}) \\
&\geq A\Big(((1 - \lambda)x + \lambda\bar{x}, (1 - \lambda)x' + \lambda\bar{x}'), W((1 - \lambda)x' + \lambda\bar{x}')\Big) \\
&\geq A\Big(((1 - \lambda)x + \lambda\bar{x}, (1 - \lambda)x' + \lambda\bar{x}'), (1 - \lambda)W(x') + \lambda W(\bar{x}')\Big) \\
&> (1 - \lambda)A((x, x'), W(x')) + \lambda A((\bar{x}, \bar{x}'), W(\bar{x}')) \\
&= (1 - \lambda)W(x) + \lambda W(\bar{x}).
\end{aligned}
$$

Consequently, W must be strictly concave. $\qquad \square$

The next lemma provides the link between the study of optimal growth paths in a strictly concave model and the theory of dynamical systems.

Lemma 3.5 *Let (T, A) be a strictly concave optimal growth model with state space $X \subseteq \mathbf{R}^n$. Then it holds for every $x \in X$ that the supremum on the right hand side of the Bellman equation (3.1) is attained by a unique vector $x' = h(x) \in T_x$. The function $h : X \mapsto X$ defined in that way is continuous and has the following property: for every $x_0 \in X$, the sequence $_0x = (x_0, x_1, x_2, \ldots)$ defined by $x_{t+1} = h(x_t)$ for all $t \in \mathbf{Z}_0^+$ is the unique optimal growth path emanating from the initial state x_0.*

PROOF. From Lemmas 3.2 and 3.4 it follows that the optimal value function W is strictly concave and continuous. This implies that the function $x' \mapsto A((x, x'), W(x'))$ from T_x to $\mathbf{R}$ is strictly concave and continuous. Because T_x is compact (see Lemma 2.1), the supremum on the right hand side of (3.1) is attained by a unique vector $x' = h(x)$. The Maximum Theorem shows that h is continuous. Now assume that $_0x \in \mathbf{F}(x_0)$ is a feasible growth path which satisfies $x_{t+1} = h(x_t)$ for all $t \in \mathbf{Z}_0^+$. From the definition of the utility functional (see Equation (2.3)) we have for all $t \in \mathbf{Z}_0^+$

$$U(_tx) = A((x_t, x_{t+1}), U(_{t+1}x)).$$

On the other hand, by the construction of the path $_0x$ and from Lemma 3.1 we also have

$$W(x_t) = A((x_t, x_{t+1}), W(x_{t+1}))$$

for all $t \in \mathbf{Z}_0^+$. From these two equations and from the Lipschitz condition (2.5) we obtain

$$|W(x_t) - U(_tx)| \leq \bar{\delta}|W(x_{t+1}) - U(_{t+1}x)|$$

for all $t \in \mathbf{Z}_0^+$. In the same way as in the proof of Lemma 3.1 we can show that this implies

$$|W(x_0) - U(_0x)| \leq \bar{\delta}^t|W(x_t) - U(_tx)| \tag{3.4}$$

for all $t \in \mathbf{Z}_0^+$. Again as in the proof of Lemma 3.1 one can see that ψ-boundedness of W and ϕ-boundedness of U (see Theorem 2.2) together with the assumptions $\bar{\delta} < 1$ and $\beta\bar{\delta} < 1$ imply that the right hand side of (3.4) converges to zero as t approaches infinity. This proves that $U(_0x) = W(x_0)$ and it follows from Definition 2.5 that $_0x$ is an optimal path.

Finally, we prove uniqueness of the optimal path $_0x$. To this end note that the optimality principle implies that $W(x_t) = U(_tx)$ for all $t \in \mathbf{Z}_0^+$. Therefore, we obtain from (2.3) and (3.1) the following:

$$\begin{aligned} \sup \left\{ A((x_t, x'), W(x')) \,\Big|\, x' \in T_{x_t} \right\} &= W(x_t) \\ &= U(_tx) \\ &= A((x_t, x_{t+1}), U(_{t+1}x)) \\ &= A((x_t, x_{t+1}), W(x_{t+1})). \end{aligned}$$

Because the supremum on the left hand side of this equation is attained uniquely at $x' = h(x_t)$, it must hold that $x_{t+1} = h(x_t)$ for all $t \in \mathbf{Z}_0^+$. This completes the proof of uniqueness. $\qquad\square$

The function h referred to in Lemma 3.5 is called the optimal policy function. For every feasible state vector x_t it determines the unique optimal successor state $x_{t+1} = h(x_t)$. In models which are not strictly concave the optimal successor state is in general not uniquely determined. In that case one has to replace the optimal policy function h by an optimal policy correspondence. Nevertheless, it is possible to define an optimal policy function also for non-concave models: simply choose an arbitrary selection from the optimal policy correspondence. To make this point clear, let us state the following definition.

Definition 3.2 Let (T, A) be an optimal growth model with state space $X \subseteq \mathbf{R}^n$. A function $h : X \mapsto X$ is called an *optimal policy function* for this model if the following is true: for every initial state $x_0 \in X$ it holds that the sequence $_0x = (x_0, x_1, x_2, \ldots)$ defined by $x_{t+1} = h(x_t)$ for all $t \in \mathbf{Z}_0^+$ is an optimal growth path.

We conclude this section by formulating necessary and sufficient optimality conditions for an optimal growth model (T, A). We would like to emphasize that these conditions are valid without any differentiability or concavity assumptions. We first prove the necessity of the conditions.

Theorem 3.1 *Let (T, A) be an optimal growth model with state space $X \subseteq \mathbf{R}^n$. If the function $h : X \mapsto X$ is an optimal policy function for this model, then there exist functions $G : T \mapsto \mathbf{R}$ and $W : X \mapsto \mathbf{R}$ such that the following three conditions are satisfied.*

i) $G(x, x') \leq 0$ for all $(x, x') \in T$ and $G(x, h(x)) = 0$ for all $x \in X$.

ii) The function W is ψ-bounded, i.e., $W \in \mathcal{B}_\psi(X)$.

iii) The equation $A((x, x'), W(x')) = G(x, x') + W(x)$ holds for all $(x, x') \in T$.

Moreover, the function W is the optimal value function of (T, A). If (T, A) is a strictly concave optimal growth model, then it follows that both the function W and the mapping $x' \mapsto G(x, x')$ are strictly concave.

PROOF. Define the function W as the optimal value function of the model (T, A) and the function G as required by condition iii) of the theorem. Then it follows from Lemma 2.8 that condition ii) is satisfied. It remains to be shown that condition i) holds. The first assertion, $G(x, x') \leq 0$, follows immediately from the definition of the function G and from Lemma 3.1. To verify $G(x, h(x)) = 0$ we denote by

$$_0x = (x, h(x), h(h(x)), \ldots)$$

the unique trajectory of h emanating from the initial state x. Since h is an optimal policy function, it follows that $_0x$ is an optimal path. Consequently we have $W(x) = U(_0x)$. Together with the monotonicity of the aggregator function with respect to its last argument we obtain

$$W(x) = U(_0x) = A((x, h(x)), U(_1x)) \leq A((x, h(x)), W(h(x)))$$

where $_1x$ is defined by $_0x = (x, _1x)$. This is equivalent to $G(x, h(x)) \geq 0$ and the proof of condition i) is therefore complete.

Now recall from Lemma 3.4 that W is a strictly concave function whenever the model (T, A) is strictly concave. Because of the monotonicity of the aggregator function A and because of the definition of G this, in turn, implies that the function $x' \mapsto G(x, x')$ is strictly concave. The proof of the theorem is now complete. $\square$

The following theorem shows that the conditions of Theorem 3.1 are also sufficient for the optimality of a given policy function h.

Theorem 3.2 *Let (T, A) be an optimal growth model with state space $X \subseteq \mathbf{R}^n$ and let $h : X \mapsto X$ be a function such that $(x, h(x)) \in T$ for all $x \in X$. If there exist functions $G : T \mapsto \mathbf{R}$ and $W : X \mapsto \mathbf{R}$ such that the conditions i) - iii) of Theorem 3.1 are satisfied, then it follows that h is an optimal policy function for the model (T, A) and that W is the optimal value function.*

PROOF. The conditions stated in part i) and part iii) of Theorem 3.1 imply that $W(x) \geq A((x, x'), W(x'))$ for all $(x, x') \in T$ and $W(x) = A((x, h(x)), W(h(x)))$ for all $x \in X$. This shows that the function W solves the Bellman equation (3.1). Since ψ-boundedness of the function W is assumed in condition ii) of Theorem 3.1, it follows from Lemma 3.1 that W must be the optimal value function of the growth model (T, A). Now let $x_0 \in X$ be an arbitrary initial state and define the feasible growth path $_0x = (x_0, x_1, x_2, \dots) \in \mathbf{F}$ recursively by $x_{t+1} = h(x_t)$ for all $t \in \mathbf{Z}_0^+$. Then we have

$$W(x_t) = A((x_t, x_{t+1}), W(x_{t+1}))$$

for all $t \in \mathbf{Z}_0^+$ by construction, and

$$U(_tx) = A((x_t, x_{t+1}), U(_{t+1}x))$$

for all $t \in \mathbf{Z}_0^+$ by the definition of the utility functional U. From this point onwards the proof that $_0x$ is an optimal growth path is identical to the corresponding section in the proof of Lemma 3.5. Since x_0 was chosen arbitrarily, it follows that h is an optimal policy function. This completes the proof. $\square$

Theorems 3.1 and 3.2 have been stated in terms of an optimal policy function. It is obvious that one can formulate these results equivalently in terms of optimal growth paths. The next corollary is such a reformulation of Theorem 3.1 which follows easily from our earlier results.

Corollary 3.1 *Let (T, A) be an optimal growth model with state space $X \subseteq \mathbf{R}^n$ and let $_0x = (x_0, x_1, x_2, \ldots)$ be an optimal growth path. Then there exist functions $G : T \mapsto \mathbf{R}$ and $W : X \mapsto \mathbf{R}$ such that the conditions ii) and iii) of Theorem 3.1 are satisfied and such that the following is true:*

i') $G(x, x') \leq 0$ for all $(x, x') \in T$ and $G(x_t, x_{t+1}) = 0$ for all $t \in \mathbf{Z}_0^+$.

Moreover, the function W is the optimal value function of (T, A) and it is strictly concave whenever (T, A) is strictly concave.

3.2 Transformations

In this section we discuss two types of transformations of optimal growth models which will be needed in the proofs of the Minimum Impatience Theorems. The first one leaves the set of optimal growth paths completely invariant whereas the second one basically performs a time transformation on the optimal growth paths. We begin with a formal definition of the equivalence of two optimal growth models.

Definition 3.3 Let (T, A) and (T, A') be two optimal growth models with a common technology T defined on the state space $X \subseteq \mathbf{R}^n$ and, hence, a common set of feasible growth paths $\mathbf{F}$. The two models are said to be *equivalent* if the following is true: a feasible growth path $_0x \in \mathbf{F}$ is an optimal path for the model (T, A) if and only if it is also an optimal path for the model (T, A').

Let (T, A) and (T, A') be two optimal growth models and denote their utility functionals by U and U', respectively. Of course, if U and U' are ordinally equivalent on the feasible set $\mathbf{F}$, then this implies that the two models are equivalent in the sense of Definition 3.3. On the other hand, ordinal equivalence of U and U' is not necessary for the equivalence of (T, A) and (T, A'). For example, it is sufficient that the utility functionals are ordinally equivalent on each of the sets $\mathbf{F}(x)$, $x \in X$, but not on the entire space $\mathbf{F}$. We are now going to introduce a method by which one can construct an optimal growth model (T, A') that is equivalent to a given model (T, A) and for which the corresponding utility functionals have exactly the property described above. It has to be mentioned that this method is nothing but the extension to the class of general recursive dynamic programming models of a construction that has already been used earlier in the additively separable case (see, for example, [13, Theorem 3]).

Lemma 3.6 *Let (T, A) be an optimal growth model with state space $X \subseteq \mathbf{R}^n$ and let U be the utility functional generated by the aggregator function A. Moreover, let $p \in \mathbf{R}^n$ be an arbitrary vector and define the function $A' : T \times \mathbf{R} \mapsto \mathbf{R}$ by*

$$A'((x, x'), z) = A((x, x'), z - p \cdot x') + p \cdot x.$$

Then it holds that A' is an aggregator function for T, that A' generates the utility functional $U'(_0x) = U(_0x) + p \cdot x_0$, and that the optimal growth models (T, A) and (T, A') are equivalent.

PROOF. First we have to show that (T, A') is an optimal growth model, i.e., that the function A' satisfies the conditions stated in Definition 2.3. To this end we define the vector $\bar{v}' \in \mathbf{R}^n$ by $\bar{v}' = \bar{v} + p$ where $\bar{v} \in \mathbf{R}^n$ is the vector mentioned in condition i) of Definition 2.3. Because of $|x_0| \le \|_0x\|_\beta$ it is easy to see that

$$|A'((x_0, x_1), \bar{v}' \cdot x_1)| = |A((x_0, x_1), \bar{v} \cdot x_1) + p \cdot x_0|$$
$$\le |A((x_0, x_1), \bar{v} \cdot x_1)| + |p| \|_0x\|_\beta.$$

It follows from this inequality and from the ϕ-boundedness of the mapping $_0x \mapsto A((x_0, x_1), \bar{v} \cdot x_1)$ that condition i) of Definition 2.3 is satisfied with A and $\bar{v}$ replaced by A' and $\bar{v}'$, respectively. Monotonicity of A' with respect to z follows immediately from the corresponding assumption on A. Finally, because the inequality

$$|A'((x, x'), z) - A'((x, x'), z')|$$
$$= |A((x, x'), z - p \cdot x') - A((x, x'), z' - p \cdot x')|$$
$$\le \bar{\delta}|z - z'|$$

holds for all $(x, x') \in T$ and for all $z \in \mathbf{R}$, we see that A' satisfies the same Lipschitz condition with respect to future utility as the function A. This proves that conditions ii) and iii) of Definition 2.3 are also satisfied by A'. Therefore, A' qualifies as an aggregator function for T.

Now define the utility functional $U' : \mathbf{F} \mapsto \mathbf{R}$ by $U'(_0x) = U(_0x) + p \cdot x_0$ for all $_0x \in \mathbf{F}$. Then we have

$$U'(_0x) = A((x_0, x_1), U(_1x)) + p \cdot x_0$$
$$= A((x_0, x_1), U'(_1x) - p \cdot x_1) + p \cdot x_0$$
$$= A'((x_0, x_1), U'(_1x))$$

which shows that U' satisfies the functional equation (2.3) with A replaced by A'. Because U' is easily seen to be ϕ-bounded, it follows from Theorem 2.2 that U' is the unique utility functional corresponding to the aggregator function A'.

Finally, for every $x \in X$, the functionals U and U' are ordinally equivalent on $\mathbf{F}(x)$ which shows that the models (T, A) and (T, A') must be equivalent. $\square$

We say that the aggregator function A' defined in the above lemma is obtained from A by a *translation*. The vector p is called the *translation vector*. Since any vector $p \in \mathbf{R}^n$ qualifies as a possible translation vector, we see immediately that the inverse problem of dynamic programming can never have a unique solution (see Section 1.1 for a discussion of the inverse problem). From any given solution (T, A) of the inverse problem one can derive uncountably many equivalent solutions (T, A') simply by translating the model (T, A) along all possible vectors p.

The method of generating an equivalent optimal growth model by translating a given aggregator function preserves many important properties of the original model. Such properties are called *translation invariant*. The most important translation invariant property is, of course, the optimality of a feasible growth path. Another translation invariant property has already been discovered in the proof of Lemma 3.6: namely, that $\bar{\delta}$ is a Lipschitz constant of the aggregator function with respect to future utility. This property as well as three others are listed in the following lemma.

Lemma 3.7 *We posit the assumptions of Lemma 3.6. Furthermore, let $D \subseteq \mathbf{F}$ be an arbitrary non-empty set of feasible paths and let $\delta \in [0, 1)$ be a real number. Each of the following properties is translation invariant, i.e., it is valid in the model (T, A) if and only if it is valid in the model (T, A'):*

i) *The optimal growth model is additively separable.*

ii) *The optimal growth model is (strictly) concave.*

iii) *The aggregator function satisfies the Lipschitz condition (2.5) with $\bar{\delta} = \delta$.*

iv) *The minimal discount factor of the aggregator function on the set D is equal to δ.*

PROOF. The proof that these properties are translation invariant is almost immediate from the definition of a translation and from Lemma 3.6. We leave it to the reader to work out the details. $\square$

The intended use of a translation is to generate an equivalent optimal growth model satisfying certain monotonicity conditions without altering other important structural properties of the original model. Two such applications of translations are described in Lemmas 3.8 and 3.9, respectively.

Lemma 3.8 *Let (T, A) be an optimal growth model with state space $X \subseteq \mathbf{R}^n$ and assume that the function $(x, x') \mapsto A((x, x'), z)$ from T to $\mathbf{R}$ satisfies a Lipschitz condition with a Lipschitz constant M which is independent of z. Moreover, assume that there exists $\underline{\delta} > 0$ such that $|A((x, x'), z) - A((x, x'), z')| \geq \underline{\delta}|z - z'|$ holds for all $(x, x') \in T$ and $z, z' \in \mathbf{R}$. Then there exists a vector $p \in \mathbf{R}^n$ such that the optimal growth model (T, A') obtained from (T, A) by a translation along the vector p has the following properties:*

i) The function $x \mapsto A'((x, x'), z)$ from $T_{x'} = \{x \in X \,|\, (x, x') \in T\}$ to $\mathbf{R}$ is strictly increasing for all $x' \in X$ for which $T_{x'}$ is non-empty and for all $z \in \mathbf{R}$.

ii) The function $x' \mapsto A'((x, x'), z)$ from T_x to $\mathbf{R}$ is strictly decreasing for all $x \in X$ and for all $z \in \mathbf{R}$.

PROOF. In the course of this proof we denote the components of a vector by subscripts, that is, a vector $x \in \mathbf{R}^n$ will be written as $x = (x_1, x_2, \ldots, x_n)$. This notation should not be confused with the time subscripts used earlier. Moreover, we shall make use of the fact that for every non-negative vector $x \in \mathbf{R}^n$ the following inequality is true:

$$\sum_{i=1}^{n} x_i \geq |x|. \tag{3.5}$$

Define the vector $p \in \mathbf{R}^n$ by $p = (k, k, \ldots, k)$ where k is any real number satisfying $k > \underline{\delta}^{-1} M$ and denote by A' the aggregator function obtained from A by a translation along the vector p. Now let $x' \in X$ be given and let x and y be two vectors in $T_{x'}$ with $x \geq y$ and $x \neq y$. Using the uniform Lipschitz constant M and (3.5) we obtain for all $z \in \mathbf{R}$

$$\begin{aligned}
A'((x, x'), z) - A'((y, x'), z) &= A((x, x'), z - p \cdot x') - A((y, x'), z - p \cdot x') \\
&\quad + p \cdot (x - y) \\
&\geq -M|x - y| + k \sum_{i=1}^{n} (x_i - y_i) \\
&\geq (k - M)|x - y| \\
&> 0.
\end{aligned}$$

This proves that $A'((x, x'), z)$ is strictly increasing with respect to x.

Now let $x \in X$ be given and let x' and y' be two vectors in T_x with $x' \geq y'$ and $x' \neq y'$. In a similar way as before taking also into account the fact that $\underline{\delta} > 0$ we obtain

$$\begin{aligned}
A'((x, x'), z) - A'((x, y'), z) &= A((x, x'), z - p \cdot x') - A((x, y'), z - p \cdot y') \\
&= A((x, x'), z - p \cdot x') - A((x, y'), z - p \cdot x')
\end{aligned}$$

$$+A((x, y'), z - p \cdot x') - A((x, y'), z - p \cdot y')$$
$$\leq M|x' - y'| - \delta p \cdot (x' - y')$$
$$\leq (M - k\underline{\delta})|x' - y'|$$
$$< 0.$$

This proves that $A'((x, x'), z)$ is strictly decreasing with respect to x'. $\qquad\square$

The monotonicity properties discussed in the above lemma have a very natural interpretation in the optimal growth context and they have been assumed to hold in a large number of models (see [63]). They simply state that a higher initial capital stock leads to higher utility and that a higher target capital stock can only be achieved at the expense of reduced utility.

Whereas Lemma 3.8 dealt with monotonicity properties of the aggregator function, our next result is about monotonicity properties of the optimal value function. To this end recall that a concave function f defined on a convex set $C \subseteq \mathbf{R}^k$ is said to be subdifferentiable at the point $x \in C$ if there exists a vector $p \in \mathbf{R}^k$ such that

$$f(x') \leq f(x) + p \cdot (x' - x)$$

holds for all $x' \in C$. The set of all vectors $p \in \mathbf{R}^k$ having this property is called the subdifferential of f at x and it is denoted by $\partial f(x)$. The vector p itself is called a subgradient. Note also that a finite concave function is subdifferentiable at every point in the relative interior of C.[5]

Lemma 3.9 *Let (T, A) be a concave optimal growth model with state space $X \subseteq \mathbf{R}^n$ and optimal value function $W : X \mapsto \mathbf{R}$. Let $Y \subseteq \mathbf{R}^n$ be an affine set such that $X \cap Y \neq \emptyset$ and denote the restriction of the function W to the set $X \cap Y$ by W_Y. If $y_0 \in X \cap Y$ is a point at which the concave function W_Y is subdifferentiable, then there exists a vector $p \in \mathbf{R}^n$ such that the concave optimal growth model (T, A') obtained from (T, A) by a translation along the vector p has the following property: the optimal value function W' of (T, A') attains its maximum over the set $X \cap Y$ at the given point y_0.*

PROOF. It follows from Lemma 3.6 that the utility functional of the translated model (T, A') is given by $U'(_0x) = U(_0x) + p \cdot x_0$ where U denotes the utility functional of the original model (T, A). Obviously, this implies that the optimal value function W' of (T, A') is given by $W'(x) = W(x) + p \cdot x$. Now choose the vector $p \in \mathbf{R}^n$ in such a way that $-p \in \partial W_Y(y_0)$. It follows immediately from this specification that $0 \in \partial W'_Y(y_0)$ where W'_Y denotes the restriction of W' to

---

[5]The relative interior of a convex set $C \subseteq \mathbf{R}^k$ is the interior of C when it is regarded as a subset of the unique smallest affine set in $\mathbf{R}^k$ containing C. The relative interior of C is non-empty whenever C is non-empty (see [56]).

the set $X \cap Y$. Because W_Y' is a concave function, the condition $0 \in \partial W_Y'(y_0)$ is sufficient for y_0 to be a maximizing point of W_Y' on its domain. This completes the proof. □

If we choose the affine set $Y = \mathbf{R}^n$, then Lemma 3.9 shows that one can translate every concave optimal growth model in such a way that its optimal value function W attains its maximum at a predetermined point y_0 in the relative interior of the state space X.

We now turn to the second transformation of optimal growth models which will become important in Chapter 4. To this end, we define the iterates of a function $h : X \mapsto X$ in the usual way, that is, $h^{(0)} = \mathrm{id}_X$ and $h^{(t+1)} = h \circ h^{(t)}$ for all $t \in \mathbf{Z}_0^+$. Our aim is to transform a given optimal growth model (T, A) with optimal policy function h in such a way that the new model (T', A') has the τ-th iterate $h^{(\tau)}$ as its optimal policy function. In other words, we are looking for a transformation of the model (T, A) which induces a time transformation on the corresponding optimal policy functions. This transformation will be called the *time-τ transformation*.

Lemma 3.10 *Let (T, A) be an optimal growth model with state space $X \subseteq \mathbf{R}^n$ and let $\tau \in \mathbf{Z}^+$ be an arbitrary positive integer. Then there exists another optimal growth model (T', A') with the same state space X such that the following conditions are true:*

  i) *If T is a β-bounded technology, then T' is a β^τ-bounded technology.*

 ii) *The globally minimal discount factor of A' is greater than or equal to the τ-th power of the globally minimal discount factor of A, that is, $\underline{\delta}(\mathbf{F}') \geq [\underline{\delta}(\mathbf{F})]^\tau$ where $\mathbf{F}$ and $\mathbf{F}'$ denote the set of feasible paths in (T, A) and (T', A'), respectively.*

iii) *If $h : X \mapsto X$ is the optimal policy function of (T, A) then $h^{(\tau)}$ is the optimal policy function of (T', A').*

 iv) *Both models (T, A) and (T', A') have the same optimal value function.*

  v) *If (T, A) is concave (strictly concave, additively separable), then the same is true for the model (T', A').*

PROOF. Let T be a β-bounded technology. We say that a capital stock $x' \in X$ is τ-reachable from another capital stock $x \in X$ if there exist feasible capital stocks $x_i \in X$, $i = 1, 2, \ldots, \tau - 1$, such that $x_1 \in T_x$, $x' \in T_{x_{\tau-1}}$, and $x_{i+1} \in T_{x_i}$ for all $i = 1, 2, \ldots, \tau - 2$. Let us define the new technology T' on the state space X by

$$T' = \Big\{ (x, x') \,\Big|\, x \in X \text{ and } x' \text{ is } \tau\text{-reachable from } x \Big\}.$$

Note that the correspondence $x \mapsto T'_x$ is the τ-th iterate of $x \mapsto T_x$, that is, $T'_x = T^{(\tau)}_x$. We leave it to the reader to verify that T' is a β^τ-bounded technology in the sense of Definition 2.1. Next we define the aggregator function $A' : T' \times \mathbf{R} \mapsto \mathbf{R}$ by

$$A'((x, x'), z) = \sup A((x, x_1), A((x_1, x_2), \ldots, A((x_{\tau-1}, x'), z) \ldots)) \tag{3.6}$$

where the supremum is taken over all finite sequences $(x_1, x_2, \ldots, x_{\tau-1})$ such that $x_1 \in T_x$, $x' \in T_{x_{\tau-1}}$, and $x_{i+1} \in T_{x_i}$ for all $i = 1, 2, \ldots, \tau - 2$. Since x' is τ-reachable from x, we know that at least one such sequence exists and, consequently, that the aggregator function A' is well defined. From the continuity of the aggregator function A and from the compactness result stated in Lemma 2.1 it follows that the supremum in (3.6) is actually a maximum. Lemma 2.5 shows that there exist a vector $\bar{v} \in \mathbf{R}^n$ and a constant $M'_1 > 0$ such that for all $(x_{\tau-1}, x') \in T$ it holds that

$$|A((x_{\tau-1}, x'), \bar{v} \cdot x')| \leq M'_1(1 + |x_{\tau-1}|).$$

Using Lemma 2.5 once again as well as the Lipschitz condition (2.5) and condition iii) of Definition 2.1 we obtain

$$
\begin{aligned}
&|A((x_{\tau-2}, x_{\tau-1}), A((x_{\tau-1}, x'), \bar{v} \cdot x'))| \\
&\quad \leq |A((x_{\tau-2}, x_{\tau-1}), \bar{v} \cdot x_{\tau-1})| + \bar{\delta}|\bar{v} \cdot x_{\tau-1} - A((x_{\tau-1}, x'), \bar{v} \cdot x')| \\
&\quad \leq M'_1(1 + |x_{\tau-2}|) + \bar{\delta}\Big[|\bar{v}||x_{\tau-1}| + M'_1(1 + |x_{\tau-1}|)\Big] \\
&\quad \leq M'_1(1 + |x_{\tau-2}|) + \bar{\delta}\Big[|\bar{v}|(\alpha + \beta|x_{\tau-2}|) + M'_1(1 + \alpha + \beta|x_{\tau-2}|)\Big] \\
&\quad \leq M'_2(1 + |x_{\tau-2}|)
\end{aligned}
$$

where $M'_2 = M'_1(1 + \bar{\delta}) + \bar{\delta}(M'_1 + |\bar{v}|) \max\{\alpha, \beta\}$. Proceeding in this way backwards, we finally obtain

$$|A'((x, x'), \bar{v} \cdot x')| \leq M'_\tau(1 + |x|)$$

for some constant $M'_\tau > 0$ which depends only on $\alpha, \beta, \bar{\delta}, |\bar{v}|,$ and M'_1. Again referring to Lemma 2.5 we can therefore conclude that the aggregator function A' satisfies condition i) of Definition 2.3.

Let $(x_1, x_2, \ldots, x_{\tau-1})$ be the sequence which attains the supremum in (3.6). Analogously, let $(x'_1, x'_2, \ldots, x'_{\tau-1})$ be the sequence which attains the supremum in (3.6) when z is replaced by z'. Then we have

$$
\begin{aligned}
&A'((x, x'), z) - A'((x, x'), z') \\
&\quad = A((x, x_1), A((x_1, x_2), \ldots, A((x_{\tau-1}, x'), z) \ldots)) \\
&\qquad - A((x, x'_1), A((x'_1, x'_2), \ldots, A((x'_{\tau-1}, x'), z') \ldots))
\end{aligned}
$$

$$\leq A((x,x_1), A((x_1,x_2),\ldots, A((x_{\tau-1},x'),z)\ldots))$$
$$-A((x,x_1), A((x_1,x_2),\ldots, A((x_{\tau-1},x'),z')\ldots))$$
$$\leq \bar{\delta}\Big|A((x_1,x_2),\ldots, A((x_{\tau-1},x'),z)\ldots)$$
$$-A((x_1,x_2),\ldots, A((x_{\tau-1},x'),z)\ldots)\Big|$$
$$\vdots$$
$$\leq \bar{\delta}^{\tau-1}\Big|A((x_{\tau-1},x'),z) - A((x_{\tau-1},x'),z')\Big|$$
$$\leq \bar{\delta}^\tau|z - z'|.$$

By interchanging the roles of z and z' we also obtain

$$A'((x,x'),z') - A'((x,x'),z) \leq \bar{\delta}^\tau|z - z'|$$

which shows that the function $z \mapsto A'((x,x'),z)$ satisfies a Lipschitz condition with Lipschitz constant $\bar{\delta}^\tau$. In a similar inductive way one can show that condition ii) of the lemma is satisfied. Monotonicity of A' with respect to z follows immediately from the corresponding monotonicity assumption on A. Taking all together, it has been shown that A' is an aggregator function for the technology T'.

Now let us denote by $W : X \mapsto X$ the optimal value function of the original model (T, A). From Lemma 3.1 we obtain

$$W(x) = \sup\Big\{ A((x,x_1), W(x_1)) \Big| x_1 \in T_x \Big\}$$
$$= \sup\Big\{ A\big((x,x_1), \sup\big\{ A((x_1,x_2), W(x_2)) \big| x_2 \in T_{x_1} \big\}\big) \Big| x_1 \in T_x \Big\}$$
$$= \sup\Big\{ A\big((x,x_1), A((x_1,x_2), W(x_2))\big) \Big| x_1 \in T_x, x_2 \in T_{x_1} \Big\}$$
$$\vdots$$
$$= \sup\Big\{ A'((x,x_\tau), W(x_\tau)) \Big| x_\tau \in T'_x \Big\}. \tag{3.7}$$

This proves that W satisfies the Bellman equation of problem (T', A'). Since W is a ψ-bounded function (by Lemma 2.8) it follows from Lemma 3.1 that W is indeed also the optimal value function of the transformed model (T', A'). Moreover, it is simple to see that the supremum in (3.7) is attained at $x_\tau = h^{(\tau)}(x)$ so that $h^{(\tau)}$ is the optimal policy function of (T', A'). The proof of condition v) is left to the reader. $\qquad\square$

It should be noted that the utility functional U' of the transformed model is given by

$$U'(_0x') = \sup\Big\{ U(_0x) \Big| _0x = (x_0,x_1,x_2,\ldots) \in \mathbf{F} \text{ and } x_{k\tau} = x'_k \text{ for all } k \in \mathbf{Z}_0^+ \Big\}$$

where $_0x' = (x'_0, x'_1, x'_2, \ldots)$. In other words, U' is obtained from U by "maximizing out" all variables x_t with $t \neq k\tau$, $k \in \mathbf{Z}_0^+$.

3.3 Indeterminacy and Turnpikes

One of the most important implications of the results presented in Section 3.1 is that the optimal growth paths in a strictly concave optimal growth model (T, A) are the trajectories of the dynamical system

$$x_{t+1} = h(x_t)$$

where the function $h : X \mapsto X$ is the optimal policy function of (T, A). This allows to draw from the rapidly increasing literature on the qualitative theory of dynamical systems when one is interested in characterizing the dynamical behavior of optimal growth paths. During the last two or three decades this theory has received considerable attention by many researchers, especially by mathematicians and physicists. More than that, the investigation of and the search for chaos or other complicated dynamical patterns has become an important topic of research throughout the natural and social sciences. By the late 1970's and the early 1980's economists have started to use non-linear deterministic dynamical systems to build models that could predict cycles and chaos as the "rational" response to reasonable economic assumptions. For a comprehensive discussion of this literature we refer the reader to the survey articles [11] and [14]. In this section, we present two of the most important results in this area and we discuss their relation to the Minimum Impatience Theorems which will be proved in Chapter 4.

Let us begin with the so-called Indeterminacy Theorem of Boldrin and Montrucchio [13]. As already indicated in Section 1.1, we can regard this result as an existence theorem for the inverse optimal growth problem.

Theorem 3.3 *Let $X \subseteq \mathbf{R}^n$ be a non-empty, convex, and compact set and let $h : X \mapsto X$ be a twice continuously differentiable function. Then there exists a real number $\delta_h \in (0, 1)$ such that for every number $\delta \in (0, \delta_h]$ the following is true: there exists an additively separable and strictly concave optimal growth model (T, A) with state space X and with an aggregator function of the form $A((x, x'), z) = V(x, x') + \delta z$ such that h is the unique optimal policy function for this model. Moreover, the function V can be chosen to be twice continuously differentiable, strictly increasing with respect to x, and strictly decreasing with respect to x'.*

PROOF. Let us specify the production technology set T, the short-run utility function $V : T \mapsto \mathbf{R}$, the optimal value function $W : X \mapsto \mathbf{R}$, and the function

$G : T \mapsto \mathbf{R}$ as follows:

$$T = X \times X,$$
$$V(x, x') = -\frac{1}{2}|x' - h(x)|^2 - \frac{\alpha}{2}|x|^2 + \frac{\alpha\delta}{2}|x'|^2,$$
$$W(x) = -\frac{\alpha}{2}|x|^2,$$
$$G(x, x') = V(x, x') - W(x) + \delta W(x').$$

Here, $\alpha > 0$ is a constant to be determined later. One can easily verify that with these specifications the conditions of Theorem 3.2 are satisfied. Strict concavity of the function V can be achieved by a suitable choice of the parameter α provided that the discount factor δ is sufficiently close to zero, say, $\delta < \delta_h$. This can be verified by calculating the Hessian matrix of V and its limit along a sequence of positive parameters (α_k, δ_k) with $\lim_{k\to\infty} \alpha_k = +\infty$ and $\limsup_{k\to\infty} \alpha_k \delta_k < 1$. Finally, the monotonicity properties stated in the theorem can be satisfied if one translates the model as described in Lemma 3.8.[6] $\square$

This result has the surprising implication that virtually every dynamical behavior is fully compatible with the standard assumption of optimal growth theory even under strict concavity requirements. The only two restrictions are the compactness of the state space and the smoothness of the policy function. As shown in [51], the latter requirement can be relaxed if one is content with a short-run utility function V which, instead of being twice continuously differentiable, is only of class $\mathcal{C}^{1+}$.[7] The proof of the Indeterminacy Theorem is constructive and can be used to build strictly concave optimal growth models with chaotic optimal growth paths. The major drawback of this result is that the construction requires, in general, extremely small discount factors. In other words, the constant δ_h mentioned in the theorem is usually very small. In this regard it should be noted that the small discount factor is only needed to render the model (T, A) strictly concave. In fact, it is clear from the proof of Theorem 3.3 that h is the unique optimal policy function of the model (T, A) for all discount factors $\delta \in (0, 1)$.

For the purpose of illustration and motivation let us consider the famous logistic map $h(x) = 4x(1 - x)$ which is probably the best known example of a chaotic dynamical system (see, e.g., [28] and [46]). All strictly concave optimal growth models presented in the literature which have this function as their optimal policy function require a discount factor around 0.01 (see [13] and [27]). Values in this parameter region correspond to a discount rate of several

---

[6]The uniform Lipschitz condition required in Lemma 3.8 follows from the continuity of the second order derivatives and from the compactness of X.

[7]A function is said to be of class $\mathcal{C}^{1+}$ if it is differentiable and has a Lipschitz continuous derivative.

1000% which is certainly unrealistic. The natural question one is therefore led to ask is whether we can do better than $\delta = 0.01$ by using more sophisticated constructions than those in [13] and [27]. The answer is positive as can be seen from the following lemma.

Lemma 3.11 *Consider the additively separable optimal growth model (T, A) with state space $X = [0, 1]$ and*

$$T = \left\{ (x, x') \,\middle|\, 0 \leq x \leq 1, 0 \leq x' \leq 4x(1 - x) \right\},$$

$$V(x, x') = 16x^4 - 32x^3 + \left(12 - \frac{1}{2\delta} \right) x^2 - 8x + \left(12\delta - \frac{1}{2} \right) (x')^2 + 2x',$$

$$A((x, x'), z) = V(x, x') + \delta z.$$

This model has the optimal policy function $h(x) = 4x(1 - x)$ for all $\delta \in (0, 1)$. Moreover, it is a strictly concave model for all discount factors $\delta \in (0, 1/24)$.

PROOF. Define the optimal value function $W : X \mapsto \mathbf{R}$ by

$$W(x) = - \left(12 + \frac{1}{2\delta} \right) x^2$$

and the function $G : T \mapsto \mathbf{R}$ by

$$G(x, x') = \left[x' - 4x(1 - x) \right] \left[2 - x' - 4x(1 - x) \right].$$

Given these specifications it is straightforward to verify that the sufficient optimality conditions of Theorem 3.2 are satisfied for the policy function $h(x) = 4x(1 - x)$. This proves the optimality of the logistic map. The verification of the strict concavity of the model (T, A) for all $\delta \in (0, 1/24)$ is a simple exercise in differential calculus. $\square$

Comparing the magnitudes of the discount factors which render this model strictly concave with those in [13] and [27], we see that we are about four times better. Note also that our model is not only additively separable in the sense of Definition 2.4 but that it satisfies the much stronger separability property

$$A((x, x'), z) = A_1(x) + A_2(x') + \delta z. \tag{3.8}$$

This property implies that the cross partial derivative $\partial^2 A((x, x'), z)/(\partial x \partial x')$ is zero on the whole domain of A. It is known (see, e.g., [8]) that the sign of this derivative determines the slope of the optimal policy function in the interior of T and that it is strongly related to the factor intensities of the underlying (non-reduced) two-sector growth model. The model of Lemma 3.11 can nevertheless have a non-monotonic optimal policy function because it is not an interior one.[8]

---

[8]Models with similar properties (no factor intensity reversal but non-monotonic optimal policy functions) are discussed in [52, 53].

The same construction as in Lemma 3.11 can also be applied to non-smooth policy functions. We illustrate this by means of the so-called tent map $h(x) = 1 - |2x - 1|$ which is topologically equivalent to the logistic map.

Lemma 3.12 *Consider the additively separable optimal growth model* (T, A) *with state space* $X = [0, 1]$ *and*

$$T = \left\{(x, x') \,\middle|\, 0 \leq x \leq 1,\, 0 \leq x' \leq 2x,\, 0 \leq x' \leq 2(1 - x)\right\},$$
$$A((x, x'), z) = \left(2 - \frac{1}{2\delta}\right) x^2 - 4x + \left(2\delta - \frac{1}{2}\right) (x')^2 + 2x' + \delta z.$$

This model has the optimal policy function $h(x) = 1 - |2x - 1|$ *for all* $\delta \in (0, 1)$. *Moreover, it is a strictly concave model for all discount factors* $\delta \in (0, 1/4)$.

PROOF. The proof is the same as the one of Lemma 3.11. The functions

$$W(x) = - \left(2 + \frac{1}{2\delta}\right) x^2$$

and

$$G(x, x') = (2x - x')(2x - 2 + x')$$

do the job. $\qquad\qquad\qquad\qquad\qquad\qquad\qquad\qquad\qquad\qquad\qquad\qquad\qquad\square$

As before we can see that the feasible values for the discount factor, $\delta < 0.25$, are much higher than those in previously published examples (see [51]). It should also be noted that the aggregator function of this model is a $\mathcal{C}^2$ function although the optimal policy function is not differentiable. Furthermore, the strong separability condition (3.8) is again satisfied.

The reader might be wondering whether we could obtain the logistic map and the tent map as optimal policy functions for relatively large discount factors only because of the particular specifications of the technology sets T. This is not the case as can be seen from the following lemma in which the logistic map turns out to be an *interior* optimal policy function for the same range of discount factors as in Lemma 3.11. Of course, we have to sacrifice the separability condition (3.8).[9]

---

[9]We would like to emphasize that the proof of Lemma 3.13 is nothing but a special case of the one for Theorem 3.3 which is obtained by setting $X = [0, 1]$, $h(x) = 4x(1 - x)$, and $\alpha = \delta^{-1} - 16$. It seems, therefore, that the construction used in Theorem 3.3 is more powerful than the one used by Boldrin and Montrucchio in their paper. As a matter of fact, it is mentioned in [13, p. 37] that their proof works only for $\delta < 0.01263 \approx 1/80$.

Lemma 3.13 *Consider the additively separable optimal growth model* (T, A) *with state space* $X = [0, 1]$, $T = X \times X$, *and*

$$A((x, x'), z) = -16x^4 + 32x^3 - x^2/\delta - 16\delta(x')^2 - 8x^2x' + 8xx' + \delta z.$$

This model has the optimal policy function $h(x) = 4x(1 - x)$ *for all* $\delta \in (0, 1)$. *Moreover, it is a strictly concave model for all discount factors* $\delta \in (0, 1/24)$.

PROOF. The proof is the same as the one of Lemma 3.11 except that we specify the functions W and G by

$$W(x) = -\left(\frac{1}{\delta} - 16\right) x^2$$

and

$$G(x, x') = -\left[x' - 4x(1 - x)\right]^2.$$

□

Encouraged by the fact that the admissible discount factors in Lemma 3.11 and Lemma 3.13 are much higher than those derived from other known examples, one might conjecture that it is possible to push the limiting value for the discount factor even closer to one by resorting to ever more sophisticated constructions. Is it perhaps possible to find, for every $\delta \in (0, 1)$, a strictly concave and additively separable optimal growth model which is solved by the logistic map? Or does there exist an upper bound $\delta^h \in (0, 1)$ such that the logistic map is certainly not the optimal policy function in any such model with a discount factor greater than δ^h? The answer to these questions will be given shortly. In fact, it will follow from the Minimum Impatience Theorems in Chapter 4 that the number $\delta^h = 0.25$ qualifies as such an upper bound for the logistic map even if we consider the general class of strictly concave optimal growth models without the restriction of additive separability. This bound is six times larger than the discount factor realized in Lemmas 3.11 and 3.13 but, nevertheless, it shows that the minimum rate of impatience required for the optimality of the logistic map is extremely high (note that $\delta = 0.25$ corresponds to a time-preference rate of 300%).

It should be clear by now what the central question is that we are trying to address in this book: for a given dynamical system h defined on a state space $X \subseteq \mathbf{R}^n$, does there exist a strictly concave optimal growth model (T, A) which admits h as its optimal policy function and for which the globally minimal discount factor $\underline{\delta}(\mathbf{F})$ exceeds a given lower bound $\delta^h < 1$? To put it differently, we would like to know if one can tell by the way a rational decision maker behaves what time-preference rate she/he has.

For some policy functions h an answer to this question will be provided by the Minimum Impatience Theorems in the following chapter. We shall demonstrate how one can find a bound δ^h such that the function h cannot be the optimal policy function in any strictly concave optimal growth model (T, A) with $\underline{\delta}(\mathbf{F}) > \delta^h$. Of course, the bound δ^h depends on the complexity of the dynamics generated by h. If h generates highly non-monotonic growth paths then δ^h is likely to be significantly smaller than one. On the other hand, it is easy to see that there are functions $h : X \mapsto X$ which can be optimal in strictly concave optimal growth models with a globally minimal discount factor which is arbitrary close to one.[10] For such a policy function a high degree of impatience is apparently not a necessary optimality condition and δ^h will be greater than or equal to one. To summarize the above discussion, we can say that a Minimum Impatience Theorem is sort of a converse to the Indeterminacy Theorem in the case that the given policy function h generates sufficiently complicated trajectories.

We conclude this section by pointing out that the results which will be proved in the following chapter do not follow from the well known Turnpike Theorem of optimal growth theory, not even in the additively separable case. To this end let us first formulate the central result from turnpike theory without precisely stating all the necessary assumptions.

Theorem 3.4 *Let (T, A) be a strictly concave optimal growth model with the additively separable aggregator function $A((x, x'), z) = V(x, x') + \delta z$. Under some technical assumptions there exists a number $\delta^m \in (0, 1)$ possibly depending on T and V such that for all discount factors $\delta \in [\delta^m, 1)$ the following is true: the optimal policy function of the given model (T, A) has a unique asymptotically stable fixpoint $\bar{x}(\delta)$. In other words, every optimal growth path $_0x = (x_0, x_1, x_2, \ldots)$ satisfies $\lim_{t \to \infty} x_t = \bar{x}(\delta)$.*

The precise statement of the conditions under which this theorem holds can be found in [59] (see also [47]). The crucial point is that the Turnpike Theorem assumes one particular optimal growth model to be given and then states that all optimal growth paths of this given model exhibit "simple" dynamics (i.e., convergence to a stationary state) provided that the discount factor is sufficiently close to one. The bound δ^m in Theorem 3.4 depends crucially on the model (T, A) under consideration (the superscript m stands for "model"). Especially the "degree of concavity" of the short-run utility function V has a

---

[10]For example, the construction used in the proof of Theorem 3.3 can be used to show that any given affine function $h : X \mapsto X$, where X is assumed to be a compact state space, is the optimal policy function in any one of a sequence of strictly concave and additively separable optimal growth models such that the corresponding sequence of discount factors converges to one from below.

decisive influence on the value of δ^m. Therefore, it is not precluded by the Turnpike Theorem that we can find a strictly concave optimal growth model with discount factor $\delta = 0.8$ which has the logistic map as its optimal policy function, another model with discount factor $\delta = 0.9$ and the same optimal policy function, a third one with $\delta = 0.95$, and so on. This possibility, however, will be ruled out by the Minimum Impatience Theorems.

Stating it briefly, the Minimum Impatience Theorems convey a similar message as the Turnpike Theorems, although viewed from a different perspective. Instead of starting with given technology and given preferences and making a statement about the optimal solutions, they assume the solution to be given and make a statement about those models which produce this solution. Both results, however, indicate that a low rate of impatience precludes the occurrence of complicated optimal growth paths.

Chapter 4

Minimum Impatience Theorems

The present chapter contains the main results of Part 1 of the book. In Section 4.1 we prove the global versions of the Minimum Impatience Theorems. One could regard these theorems simply as new necessary optimality conditions for strictly concave optimal growth models. There is, however, a very crucial feature of the Minimum Impatience Theorems which distinguishes them from other necessary optimality conditions. This feature is that they do not depend on the "givens" (T, A) of the problem except for the state space X and the minimal discount factor $\underline{\delta}(D)$ on a suitably chosen set $D \subseteq \mathbf{F}$. The way the Minimum Impatience Theorems and their corollaries can be applied is therefore quite different from the way other necessary optimality conditions are commonly used. Whereas the Euler equations or the Bellman equation are used to determine all optimal solutions to a given problem, the Minimum Impatience Theorems can be used to preclude the optimality of a given solution in a whole class of problems. In other words, these theorems represent conditions which have to be satisfied by any strictly concave optimal growth problem which admits a given function $h : X \mapsto X$ as its optimal policy function or a given sequence x_0, x_1, x_2, ... as an optimal growth path. It is clear from this description that the Minimum Impatience Theorems are more closely related to the inverse optimal growth problem than to the optimal growth problem itself.

We begin with a result (Theorem 4.1) which basically shows that a highly non-monotonic optimal growth path cannot be optimal in models with low time-preference rates. The first result of this type was proved in [62] for the additively separable case and for bounded growth paths. Here, we present a version for the general model without the assumption of additive separability. Neither do we need the boundedness assumption on the growth paths because it is replaced by the uniform summability mentioned at the end of Section 2.2. In our second main result (Theorem 4.4) we consider two different feasible growth paths and show that they cannot diverge from each other at a rate exceeding the discount rate. In other words, the driving force behind this Minimum

Impatience Theorem is not the non-monotonicity of an optimal growth path but its instability or sensitivity with respect to initial conditions.

Section 4.2 considers the one-dimensional case in more detail and derives some simple corollaries from the main theorem. For example, we show that an upper bound on the minimal discount factor in one-dimensional problems can be calculated by solving a certain equation which depends in a simple way on the given growth path. We illustrate these results by applying them to the logistic map and to the tent map. We also discuss the implications which can be derived from the Minimum Impatience Theorems in the case that a period-three cycle is an optimal growth path.

In Section 4.3 we present properties of the so-called average correspondence, which is all-important for those Minimum Impatience Theorems which are based on the non-monotonicity of optimal growth paths. In particular, we derive a representation of the average correspondence which is extremely useful to understand its local behavior at fixpoints of the policy function. These results are used in Section 4.4 where we introduce the notion of a δ-expanding stationary state of the policy function and where we prove that the existence of such a stationary state already implies a minimum degree of impatience on the side of the decision maker. Using the time-τ transformation introduced in Lemma 3.10 we can generalize this idea to obtain a Minimum Impatience Theorem based on the existence of a δ-expanding periodic point of arbitrary period τ. All these theorems use only local properties of the given optimal policy function h. Because δ-expansiveness is a rather strong property, we also show how the absolute values of the real eigenvalues of the Jacobian of h evaluated at a (not necessarily δ-expanding) fixpoint give rise to another lower bound on the time-preference rate.

Section 4.5 concludes the first part of the book by discussing several directions in which our results could perhaps be extended or generalized. In particular, we explain what would be needed to make a Minimum Impatience Theorem a true necessary optimality condition for chaos.

4.1 Main Results

In this section we prove the global Minimum Impatience Theorems for strictly concave optimal growth models. We shall present two conceptually different approaches: the first one assumes that only a single optimal growth path is given, whereas the second one requires knowledge of two paths.

It will be assumed throughout this section that $X \subseteq \mathbf{R}^n$ is a fixed state space, i.e., a non-empty, closed, and convex subset of the n-dimensional Euclidean space. Moreover, let us assume that a sequence $_0x = (x_0, x_1, x_2, \ldots)$ is given such that $x_t \in X$ holds for all $t \in \mathbf{Z}_0^+$. We are trying to answer the question of whether this sequence $_0x$ represents an optimal growth path in some strictly concave optimal growth problem (T, A) with state space X. To this end we define for every $\delta \in (0, 1]$ the set $\tilde{H}(_0x, \delta) \subseteq X$ as the set of all vectors $y \in \mathbf{R}^n$ for which there exists a sequence $\mu_1, \mu_2, \mu_3, \ldots$ of real numbers such that the following conditions are satisfied:

$$\mu_t \geq \mu_{t+1} \geq 0 \text{ for all } t \in \mathbf{Z}^+ \tag{4.1}$$

$$\sum_{t=1}^{\infty} \mu_t \delta^t = 1 \tag{4.2}$$

$$\sum_{t=1}^{\infty} \mu_t \delta^t x_t = y \tag{4.3}$$

It is easily seen that every $y \in \tilde{H}(_0x, \delta)$ is a convex combination of the vectors $x_1, x_2, x_3, \ldots$. Since X is assumed to be convex, this implies that $\tilde{H}(_0x, \delta)$ is a subset of X. Moreover, it is obvious that $\tilde{H}(_0x, \delta)$ itself is also a convex set. Finally, we have the following monotonicity result.

Lemma 4.1 *Let $_0x = (x_0, x_1, x_2, \ldots)$ be a sequence of vectors $x_t \in X$ and let δ and δ' be two real numbers such that $0 < \delta \leq \delta' \leq 1$. Then it holds that $\tilde{H}(_0x, \delta) \subseteq \tilde{H}(_0x, \delta')$.*

PROOF. Assume that $y \in \tilde{H}(_0x, \delta)$. According to the definition of the set $\tilde{H}(_0x, \delta)$ there exist real numbers μ_t, $t \in \mathbf{Z}^+$, such that conditions (4.1) - (4.3) are satisfied. Defining $\mu'_t = \mu_t(\delta/\delta')^t$ for all $t \in \mathbf{Z}^+$ it can easily be verified that these conditions remain true when δ and μ_t are replaced by δ' and μ'_t, respectively. Therefore, it follows that $y \in \tilde{H}(_0x, \delta')$ and the proof of the lemma is complete. $\square$

Before we can state the first main result of this section we need one more technical lemma.

Lemma 4.2 *Let $_0x = (x_0, x_1, x_2, \ldots)$ be an optimal growth path in some strictly concave optimal growth model (T, A) with state space X and assume $x_0 \neq x_1$. If there exists $\delta \in (0, 1]$ such that $x_0 \in \tilde{H}(_0x, \delta)$, then it holds that $x_1 \neq x_2$.*

PROOF. By strict concavity of (T, A) it follows from Lemma 3.5 that there exists an optimal policy function $h : X \mapsto X$ such that $x_{t+1} = h(x_t)$ holds for all $t \in \mathbf{Z}_0^+$. Now assume that $x_1 = x_2$. This implies that x_1 is a fixpoint of h and, consequently, that $x_t = x_1$ for all $t \in \mathbf{Z}^+$. From (4.2) and (4.3) it follows then that $\tilde{H}(_0x, \delta) = \{x_1\}$ which shows that the assumptions $x_0 \neq x_1$ and $x_0 \in \tilde{H}(_0x, \delta)$ are mutually excluding. This contradiction proves the lemma. $\square$

We are now in a position to present our first Minimum Impatience Theorem. As already mentioned it is a generalization of Theorem 1 in [62].

Theorem 4.1 *Let $X \subseteq \mathbf{R}^n$ be a non-empty, closed, and convex set and let a sequence $_0x = (x_0, x_1, x_2, \ldots)$ of vectors be given such that $x_t \in X$ holds for all $t \in \mathbf{Z}_0^+$. Assume that $x_0 \neq x_1$ and denote by D the singleton set $D = \{_0x\}$. If $_0x$ is an optimal growth path in a strictly concave optimal growth model (T, A) with state space X, then it holds that $x_0 \notin \tilde{H}(_0x, \delta)$ for all $\delta \leq \underline{\delta}(D)$ where $\underline{\delta}(D)$ is the minimal discount factor of A along the given path $_0x$.*

PROOF. Because of Lemma 4.1 it is sufficient to show that $x_0 \notin \tilde{H}(_0x, \delta)$ for $\delta = \underline{\delta}(D)$. The proof is by contradiction. Assuming that $x_0 \in \tilde{H}(_0x, \delta)$ it follows that there exists a sequence μ_1, μ_2, μ_3, $\ldots$ of real numbers such that conditions (4.1) and (4.2) hold and such that

$$\sum_{t=1}^{\infty} \mu_t \delta^t x_t = x_0. \tag{4.4}$$

Since $_0x$ is assumed to be an optimal path in the model (T, A), there must exist constants $\beta > 1$ and $\bar{\delta} \in (0, 1)$ such that the conditions $_0x \in \mathbf{S}_\beta$, $\delta = \underline{\delta}(D) \leq \bar{\delta}$, and $\beta\bar{\delta} < 1$ are satisfied. Let μ_0 be a real number such that $\mu_0 \geq \mu_1$ and define $\lambda_t = \mu_t \delta^t / (1 + \mu_0)$ for all $t \in \mathbf{Z}_0^+$. Moreover, define $\Delta = \{t \mid \lambda_t > 0\}$. One can easily verify that these specifications imply $0 < \lambda_t < 1$ for all $t \in \Delta$ as well as

$$\sum_{t \in \Delta} \lambda_t = 1 \tag{4.5}$$

and

$$\sum_{t \in \Delta} \lambda_t x_t = x_0. \tag{4.6}$$

Now we consider the infinite series

$$\sum_{t=0}^{\infty} \lambda_t x_{t+1}.$$

Because of $0 \leq \lambda_t \leq \delta^t \leq \bar{\delta}^t$, $_0x \in \mathbf{S}_\beta$, and $\beta\bar{\delta} < 1$ it follows that this series is absolutely convergent. This shows that the vector

$$y_0 = \sum_{t \in \Delta} \lambda_t x_{t+1} \tag{4.7}$$

is well defined. Let $Y \subseteq \mathbf{R}^n$ be the affine hull of the set $\{x_{t+1} \mid t \in \Delta\}$.[1] We claim that y_0 is in the relative interior of $X \cap Y$. To prove this we first note that (4.1) and (4.2) imply that $\mu_1 > 0$ which, in turn, shows that $\{0, 1\} \subseteq \Delta$. Because

---

[1] The affine hull of a non-empty set $C \subseteq \mathbf{R}^n$ is defined as the smallest affine set containing C.

58

of Lemma 4.2 we can see that the set $\{x_{t+1} \mid t \in \Delta\} \subseteq X \cap Y$ has cardinality greater than or equal to two, and that y_0 is a proper convex combination of the points in this set.[2] It is easy to show that this implies that y_0 is in the relative interior of $X \cap Y$. It follows that the restriction of the optimal value function of (T, A) to the set $X \cap Y$ is subdifferentiable at y_0 (see [56, Theorem 23.4]). In view of Lemma 3.9 and Lemma 3.7 we may therefore assume that the optimal value function W of (T, A) satisfies

$$W(y_0) \geq W(x_{t+1}) \tag{4.8}$$

for all $t \in \Delta$. Equations (4.6) and (4.7) imply that the vector (x_0, y_0) is a proper convex combination of the vectors (x_t, x_{t+1}), $t \in \Delta$. Since $_0x$ is a feasible path in (T, A), we must have $(x_t, x_{t+1}) \in T$ for all $t \in \mathbf{Z}_0^+$. Strict concavity of the model (T, A) implies that $(x_0, y_0) \in T$ and that

$$A((x_0, y_0), W(y_0)) > \sum_{t \in \Delta} \lambda_t A((x_t, x_{t+1}), W(y_0)).$$

From Corollary 3.1 it follows that there exists a non-positive function $G : T \mapsto \mathbf{R}$ such that the above inequality can be rewritten as follows:

$$G(x_0, y_0) + W(x_0)$$
$$> \sum_{t \in \Delta} \lambda_t \Big[G(x_t, x_{t+1}) + W(x_t)$$
$$+ A((x_t, x_{t+1}), W(y_0)) - A((x_t, x_{t+1}), W(x_{t+1})) \Big]. \tag{4.9}$$

Corollary 3.1 also shows that $G(x_0, y_0) \leq 0$ and that $G(x_t, x_{t+1}) = 0$ for all $t \in \mathbf{Z}_0^+$. Therefore, a necessary condition for Inequality (4.9) to hold is that

$$W(x_0) > \sum_{t \in \Delta} \lambda_t \Big[W(x_t) + A((x_t, x_{t+1}), W(y_0)) - A((x_t, x_{t+1}), W(x_{t+1})) \Big]. \tag{4.10}$$

From the definition of the minimal discount factor $\delta = \underline{\delta}(D)$ (Definition 2.7) and from (4.8) it follows that for all $t \in \Delta$ the following is true:

$$A((x_t, x_{t+1}), W(y_0)) - A((x_t, x_{t+1}), W(x_{t+1})) \geq \delta \Big[W(y_0) - W(x_{t+1}) \Big].$$

If we substitute this into (4.10) we get

$$W(x_0) > \sum_{t \in \Delta} \lambda_t \Big[W(x_t) - \delta W(x_{t+1}) + \delta W(y_0) \Big].$$

[2]A convex combination $\sum_s \lambda_s y_s$ is said to be proper if all weights λ_s are positive.

Using (4.5) and $\lambda_t = 0$ for all $t \notin \Delta$ we can rearrange the terms in this inequality to obtain[3]

$$W(y_0) < \frac{1 - \lambda_0}{\delta} W(x_0) + \sum_{t=1}^{\infty} \frac{\delta \lambda_{t-1} - \lambda_t}{\delta} W(x_t). \qquad (4.11)$$

Finally, we substitute the definition of the weights λ_t into (4.11) which yields

$$W(y_0) < \frac{1}{\delta(1 + \mu_0)} W(x_0) + \sum_{t=1}^{\infty} \frac{\mu_{t-1} - \mu_t}{1 + \mu_0} \delta^{t-1} W(x_t). \qquad (4.12)$$

On the other hand, we have

$$y_0 = \sum_{t=1}^{\infty} \lambda_{t-1} x_t$$

$$= \frac{1}{\delta(1 + \mu_0)} \sum_{t=1}^{\infty} \mu_{t-1} \delta^t x_t$$

$$= \frac{1}{\delta(1 + \mu_0)} \left[\sum_{t=1}^{\infty} \mu_t \delta^t x_t + \sum_{t=1}^{\infty} (\mu_{t-1} - \mu_t) \delta^t x_t \right].$$

From (4.4) it follows that the first sum on the right hand side is equal to x_0. Therefore, we obtain

$$y_0 = \frac{1}{\delta(1 + \mu_0)} x_0 + \sum_{t=1}^{\infty} \frac{\mu_{t-1} - \mu_t}{1 + \mu_0} \delta^{t-1} x_t.$$

Using (4.1) and (4.2) it is straightforward to verify that this is a representation of the vector y_0 as a convex combination of the vectors x_t, $t \in \mathbf{Z}_0^+$. Because W is a concave function this implies that

$$W(y_0) \geq \frac{1}{\delta(1 + \mu_0)} W(x_0) + \sum_{t=1}^{\infty} \frac{\mu_{t-1} - \mu_t}{1 + \mu_0} \delta^{t-1} W(x_t).$$

This is an obvious contradiction to (4.12) and the theorem is proven. □

In the following, we discuss various points concerning the assumptions and the application of Theorem 4.1. First of all, however, let us make a remark on the proof. It is interesting to note that this indirect proof does not involve the

---

[3]Rearranging the terms is feasible because all infinite sequences are guaranteed to converge absolutely. This follows from $\lambda_t \leq \delta^t \leq \bar{\delta}^t$, the ψ-boundedness of W (see Lemma 3.1), and the condition $\beta \bar{\delta} < 1$. The reader should compare this argument with the remark made at the end of Section 2.2.

construction of a new path $_0\tilde{x} \in \mathbf{F}(x_0)$ which yields a higher utility than the given path $_0x$ in any strictly concave optimal growth model on the state space X. In fact, optimality considerations enter the proof of Theorem 4.1 only via the necessary optimality conditions of Corollary 3.1. This becomes even more evident from the fact that the proof does not use the property that $_0x$ is the trajectory of a dynamical system (Lemma 3.5). What then is the crux of the proof? It is the interplay of the geometric properties of the aggregator function and the optimal value function, respectively, as highlighted by Corollary 3.1. To further illustrate this point consider the time-additive case in which condition iii) of Theorem 3.1 can be written as

$$V(x, x') = G(x, x') + W(x) - \delta W(x'). \tag{4.13}$$

A close look at condition i) of Theorem 3.1 makes clear that the function G can be concave only if the candidate path $_0x$ is generated by a linear policy function, that is, if $x_{t+1} = Qx_t + c$ for some constant matrix Q and some constant vector c. In this case Theorem 4.1 will not be useful. In the presence of the slightest non-linearity, however, G can no longer be a concave function. The short-run utility function V, on the other hand, is required to be concave. We can achieve the concavity of V only if we add a "sufficiently concave" term $W(x)$ on the right hand side of (4.13). This will remedy the non-concavity with respect to x but, at the same time, it will create some non-concavity with respect to x' as the term $\delta W(x')$ is subtracted. Consequently, the function $\delta W(x')$ must not be "too concave". These two conflicting goals can be achieved simultaneously only if the discount factor is sufficiently small. This is the essence of the proof of the Minimum Impatience Theorem 4.1.

Leaving the mathematical aspects aside, we now turn to the contents of Theorem 4.1. First, we note that there is a very simple interpretation of the result. In fact, scrutinizing conditions (4.1) - (4.3) reveals that every vector $y \in \tilde{H}(_0x, \delta)$ can be represented as a weighted average of the points x_t, $t \in \mathbf{Z}^+$, where the weights $\bar{\mu}_t = \delta^t \mu_t$ converge at least linearly to zero with a convergence factor δ, i.e., $\bar{\mu}_{t+1} \leq \delta\bar{\mu}_t$. This shows that Theorem 4.1 can be reformulated as follows: The initial capital stock of any non-stationary optimal growth path in a strictly concave optimal growth model cannot be represented as the weighted average of all future capital stocks, when the weights are constrained to converge at least linearly to zero with a convergence factor smaller than or equal to the minimal discount factor along the path. The intuitive explanation of this result is easy to derive. In fact, at various places in the literature the necessity of heavy discounting for the optimality of fluctuating growth paths has already been noted. We quote here from a recent article by Boldrin and Deneckere [12, p. 629] who state that "because of the strict concavity of the production possibility frontier this wandering of x_t will imply huge variations in relative prices and rates of return. A high level of discounting is then needed in order to eliminate the arbitrage possibilities that would otherwise emerge". Theorem 4.1 can be regarded as a theoretical underpinning of this statement.[4] If x_0 is an element of the set $\tilde{H}(_0x, \delta)$, then it follows that the variations in x_t are so strong that they allow for arbitrage in any strictly concave model with $\underline{\delta}(\{_0x\}) \geq \delta$. We can therefore interpret our result as a (necessary) no-arbitrage condition which has the important property that it is almost independent of the given model (T, A).

Let us look once more at the condition $x_0 \in \tilde{H}(_0x, \delta)$. We have already noted that this condition implies that x_0 is a convex combination of the points x_t, $t \in \mathbf{Z}^+$. In particular, this shows that the result stated in Theorem 4.1 is trivial if $_0x$ is a strictly monotonic sequence in some direction $a \in \mathbf{R}^n$, i.e., if the inequality $a \cdot (x_t - x_{t+1}) > 0$ holds for all $t \in \mathbf{Z}_0^+$. In that case, it is easy to see that one can separate the initial state x_0 from the set $\tilde{H}(_0x, \delta)$ by a hyperplane which is normal to the vector a. Therefore, we can conclude that Theorem 4.1 makes a non-trivial statement only if the trajectory under investigation exhibits a sufficiently high degree of non-monotonicity "in every possible direction". As a matter of fact, one could say that the smallest value of δ for which the relation $x_0 \in \tilde{H}(_0x, \delta)$ holds provides a measure of the monotonicity of $_0x$: the lower it is the less monotonic is $_0x$. With this understanding Theorem 4.1 states that the minimal discount factor along an optimal path cannot exceed its measure of monotonicity.

We have motivated the Minimum Impatience Theorems in Section 3.3 by discussing various attempts in the literature to develop optimal growth models which admit the logistic map as their optimal policy function. In the formulation of Theorem 4.1, however, we have only assumed that one single growth path $_0x$ is given which, by the way, need not even be a trajectory of a dynamical system. If we happen to know an optimal policy function h, then we can derive a lower bound on the maximal rate of impatience (i.e., an upper bound on the minimal discount factor) by first computing various trajectories of the dynamical system

$$x_{t+1} = h(x_t) \tag{4.14}$$

and then checking whether there exists a number $\delta \in (0, 1)$ such that the condition $x_0 \in \tilde{H}(_0x, \delta)$ holds true for at least one of these trajectories. For later reference, especially for the analysis presented in Section 4.4, it will be convenient to reformulate Theorem 4.1 under the assumption that a policy function $h : X \mapsto X$ is given. To this end, recall that we denote by $h^{(0)}, h^{(1)}, h^{(2)}, \ldots$

---

[4]The prices mentioned in this statement are the competitive equilibrium prices. Although a general duality theory for recursive utility maximization is not known to the author, one can conjecture by analogy to the time-additive case that the dynamical evolution of the prices (dual variables) is topologically equivalent to the one of the capital stocks (see, e.g., [48, p. 86]).

the iterates of the mapping h. Moreover, we introduce the notion of the average correspondence H associated with the given function h. It will be shown that this correspondence contains the entire information which is necessary to apply Theorem 4.1.

Definition 4.1 Let $x \in X$ be a feasible state vector and let $\delta \in (0,1]$ be a real number. Then we define $H(x,\delta) = \tilde{H}({}_0x,\delta)$ where ${}_0x = (x, h(x), h^{(2)}(x), \ldots)$ is the unique trajectory of (4.14) emanating from the initial state x. More specifically, $H(x,\delta)$ consists of all vectors $y \in X$ for which there exists a sequence of real numbers $\mu_1, \mu_2, \mu_3, \ldots$ such that conditions (4.1) and (4.2) as well as

$$\sum_{t=1}^{\infty} \mu_t \delta^t h^{(t)}(x) = y \tag{4.15}$$

are satisfied. The multifunction $H : X \times (0,1] \mapsto 2^X$ will be called the *average correspondence*.

Justification for the choice of the term "average correspondence" has already been provided earlier. Various properties of the correspondence H will be discussed in Section 4.3. Before we can present the announced reformulation of Theorem 4.1 we have to explain what we mean by a non-trivial fixpoint of the average correspondence.

Definition 4.2 Let $H : X \times (0,1] \mapsto 2^X$ be the average correspondence associated with a given mapping $h : X \mapsto X$. We say that the vector $x \in X$ is a *non-trivial fixpoint* of H at δ if the conditions $x \neq h(x)$ and $x \in H(x,\delta)$ are satisfied.

The following result is an easy corollary of Theorem 4.1 and the above definitions. Nevertheless, we state it in the form of a theorem because it is of central importance in this book.

Theorem 4.2 *Let $X \subseteq \mathbf{R}^n$ be a non-empty, closed, and convex set and let an arbitrary mapping $h : X \mapsto X$ be given. Assume that there exist $x \in X$ and $\delta \in (0,1]$ such that x is a non-trivial fixpoint of the average correspondence H at δ. If one defines D as the set consisting of the unique trajectory of h which emanates from x, i.e., $D = \{(x, h(x), h^{(2)}(x), \ldots)\}$, then there does not exist any strictly concave optimal growth model (T, A) with state space X such that h is the optimal policy function of this model and such that the minimal discount factor of A on D satisfies $\underline{\delta}(D) \geq \delta$.*

Reformulating the theorem in this way makes clear why it is called a Minimum Impatience Theorem: it states that certain policy functions (those for which there exists a non-trivial fixpoint of the associated average correspondence) require a minimum degree of impatience, i.e., a sufficiently low discount factor in order to become optimal in at least one strictly concave optimal growth model. This immediately raises the question of which properties of the given function h make the average correspondence H have a non-trivial fixpoint. We shall try to give at least some partial answers to this question in the following sections.

As compared to Theorem 4.1 the formulation of Theorem 4.2 highlights the way in which one will presumably apply the Minimum Impatience Theorems. Unlike other optimality conditions (for example, Theorem 3.1) the results derived in this chapter will not be used to determine or to characterize the set of optimal growth paths for a given model (T, A) but rather to characterize the class of optimal growth models (T, A) which admit the given function h as their optimal policy function. This shows that, properly speaking, a Minimum Impatience Theorem is a necessary condition for the existence of a solution to the inverse optimal growth problem which consists in finding all models having a given optimal policy function or a given optimal growth path.[5]

Now suppose that we want to find the best bound on the discount factor which can be obtained from Theorem 4.2 for a given mapping $h : X \mapsto X$. It is obvious that we have to solve the following non-linear mathematical programming problem:

$$\text{Minimize } \delta$$
$$\text{subject to } x \in X, \ x \neq h(x), \text{ and } x \in H(x, \delta).$$

For the time being let us call this problem P. Substituting the definition of $H(x, \delta)$ into the third constraint of P we see that there are countably many decision variables to be determined: δ, x, and μ_t, for all $t \in \mathbf{Z}^+$. This shows that problem P is an infinite dimensional programming problem. Moreover, it is highly non-linear and non-convex. Therefore, it does not seem to be easy to solve problem P exactly and approximation techniques become important.

There is also another difficulty hidden in the third constraint of P, namely, that this constraint involves all (infinitely many) iterates $h^{(t)}(x)$, $t \in \mathbf{Z}_0^+$. Especially for chaotic maps which suffer from sensitive dependence on initial conditions it is in general not possible to compute all of these iterates within sufficient accuracy. Because of this reason, let us assume that we know only a finite number of iterates $h^{(t)}(x)$, say, from $t = 1$ to $t = m$. In order to make sure that the condition $x \in H(x, \delta)$ is satisfied even without knowing the entire trajectory we have to set $\mu_t = 0$ for all $t > m$. What remains is a finite dimensional optimization problem P_m which, however, is still non-linear and non-convex. If

---

[5]See, e.g., Kurz [41] or Chang [21] for an analysis of the inverse optimal growth problem and Section 1.1 for a brief discussion.

the mapping h actually exhibits sensitive dependence on initial conditions, it seems intuitively plausible that there are many local optima for this problem.[6] Consequently, one cannot expect that algorithms based on necessary optimality conditions or descent methods would yield good results. Probabilistic methods like simulated annealing or genetic algorithms are more likely to perform well.

Anticipating the results of Section 4.3 we would like to mention that computing only finitely many iterates $h^{(t)}(x)$, $t = 1, 2, \ldots, m$, corresponds to an approximation of the average correspondence H by an increasing sequence of correspondences H_m, $m \in \mathbf{Z}^+$, such that

$$H(x, \delta) = \overline{\bigcup_{m \in \mathbf{Z}^+} H_m(x, \delta)},$$

where the bar over a set denotes its topological closure. This approach will turn out to be particularly useful for the theoretical analysis of the local properties of H, which will be needed to prove a local Minimum Impatience Theorem in Section 4.4.

One of the rare instances in which it is actually possible to compute all iterates $h^{(t)}(x)$ even for a chaotic map is when the initial state x belongs to the set of eventually periodic or eventually fixed points of the mapping h. Our numerical experiments and the examples presented in Section 4.2 seem to indicate that these points are very likely to yield good bounds in Theorem 4.1.

Finally, we should remark that it can be advantageous to apply Theorems 4.1 and 4.2 to a higher iterate $h^{(\tau)}$ of the given policy function h. Once we have found a bound on the minimum rate of impatience required for the optimality of $h^{(\tau)}$, we can also calculate such a bound for the original mapping h by application of Lemma 3.10. This approach is made precise in the following theorem.

Theorem 4.3 *Let $X \subseteq \mathbf{R}^n$ be a non-empty, closed, and convex set, $\tau \in \mathbf{Z}^+$ an arbitrary number, and $h : X \mapsto X$ given mapping. Denote by H' the average correspondence associated to the τ-th iterate $h^{(\tau)}$ of h. Assume that there exist $x \in X$ and $\delta \in (0, 1]$ such that x is a non-trivial fixpoint of H' at δ^τ. Then there does not exist any strictly concave optimal growth model (T, A) with state space X such that h is the optimal policy function of this model and such that the globally minimal discount factor of A satisfies $\underline{\delta}(\mathbf{F}) \geq \delta$.*

PROOF. The proof is by contradiction. Suppose that there is a model (T, A) with optimal policy function h and $\underline{\delta}(\mathbf{F}) \geq \delta$. By Lemma 3.10 there exists a strictly concave optimal growth model (T', A') such that $h^{(\tau)}$ is the optimal

---

[6]This conjecture is confirmed by our numerical results. See Figures 4.1 - 4.4 at the end of Section 4.2.

policy function of this model and such that $\underline{\delta}(\mathbf{F}') \geq [\underline{\delta}(\mathbf{F})]^\tau \geq \delta^\tau$. Since x is a non-trivial fixpoint of H' at δ^τ, this is a contradiction to Theorem 4.2. $\qquad\square$

We conclude the discussion of Theorems 4.1 - 4.3 by two examples. The first one is the Hénon map

$$(x, y) \mapsto (y, 1 + 0.3x - 1.4y^2) \qquad (4.16)$$

which has been analysed extensively in the literature on chaotic dynamical systems (see, e.g., [25, 34]). It has been reported in [14, p. 201] that this map is the optimal policy function of a strictly concave and additively separable optimal growth model with a discount factor $\delta = 0.2$. Let us see what we can get from Theorem 4.2 in this case. By a very simple and crude Monte Carlo search we have found that the point $(x, y) = (0.59, 0.58)$ is a non-trivial fixpoint of the average correspondence at $\delta = 0.555$. We have chosen $\mu_t = 0$ for all $t \geq 8$ so that only the iterates up to $m = 7$ had to be computed. The iterates x_t and y_t as well as the weights μ_t are listed below (rounded to five decimal places):

t	x_t	y_t	μ_t
0	0.59000	0.58000	–
1	0.58000	0.70604	0.81701
2	0.70604	0.47611	0.81614
3	0.47611	0.89446	0.81104
4	0.89446	0.02276	0.81104
5	0.02276	1.26761	0.81104
6	1.26761	−1.24275	0.81104
7	−1.24275	−0.78192	0.81104

The aforementioned bound, $\underline{\delta}(\mathbf{F}) < 0.555$, was obtained by applying Theorem 4.2 directly to the Hénon map (4.16). The bound can be improved by considering higher iterates of this mapping. For example, the point $(-0.36, 0.92)$ is a non-trivial fixpoint of the second iterate of the Hénon map at $\delta = 0.224$. Application of Theorem 4.3 with $\tau = 2$ yields $\underline{\delta}(\mathbf{F}) < 0.474$ which shows that the Hénon map requires a minimum discount rate of roughly 110%. This is already rather high so that we did not search for a better bound by applying

Theorem 4.3 with $\tau \geq 3.$[7] The data to derive $\underline{\delta}(\mathbf{F}) < 0.474$ are as follows ($\mu_t = 0$ for all $t \geq 5$):

t	x_t	y_t	μ_t
0	-0.36000	0.92000	$-$
1	-0.29296	1.15484	3.47847
2	-0.95825	0.06120	3.47690
3	0.70728	0.31802	3.47690
4	1.07059	-0.50924	2.89478

Our second example is the following two-dimensional system:

$$(x,y) \mapsto \left((x+y)/10 + 10e^{-10(x+y)}, x\right) \tag{4.17}$$

All trajectories become trapped in the region $[0,3] \times [0,3]$ after finitely many iterations so that $X = [0,3] \times [0,3]$ can be taken as the state space. The point $(x,y) = (0.47, 0.26)$ is a non-trivial fixpoint of the associated average correspondence at $\delta = 0.721$. To verify this it is enough to compute four iterations of the map and to set $\mu_t = 0$ for all $t \geq 5$. The relevant data are as follows:

t	x_t	y_t	μ_t
0	0.47000	0.26000	$-$
1	0.07976	0.47000	0.61820
2	0.09594	0.07976	0.61820
3	1.74321	0.09594	0.61732
4	0.18392	1.74321	0.00569

If we consider the second iterate of the mapping (4.17), we can see that the point $(0.3, 0.2)$ is a non-trivial fixpoint at $\delta = 0.358$ which according to Theorem 4.3

---

[7]An even better bound for this example will be derived from Theorem 4.4 below.

(with $\tau = 2$) implies that $\underline{\delta}(\mathbf{F}) < 0.599$. To verify this bound we needed only three iterations of (4.17), i.e., $\mu_t = 0$ for all $t \geq 4$. The numerical results for this example are summarized in the following table:

t	x_t	y_t	μ_t
0	0.30000	0.20000	–
1	0.19568	0.11738	1.88162
2	0.07947	0.46824	1.87700
3	1.73709	0.09658	1.87034

Non-monotonicity as dealt with by Theorems 4.1 - 4.3 is only one characteristic feature of complicated dynamics. Another one is sensitive dependence on initial conditions. In the remainder of this section we shall prove Minimum Impatience Theorems which focus on this aspect of complexity. The idea on which these results are based is the same as the one used in an unpublished paper by Hewage and Neumann [35]. More specifically, what we present here is a generalized and corrected version of part of that paper. Our results are more general because they apply to the n-dimensional case with non-additive preferences whereas [35] deals only with the one-dimensional and additively separable case. There are also some minor errors in the statements of the results in [35] which we shall point out to the reader as we go along. We begin with a technical lemma stating a certain property of strictly concave functions.

Lemma 4.3 Let $X \subseteq \mathbf{R}^n$ be a non-empty convex set, $W : X \mapsto \mathbf{R}$ a strictly concave function defined on X, $\bar{x}$ a given point in the relative interior of X, and $\delta \in (0,1)$ a real number. For every non-zero vector $y \in \mathbf{R}^n$ which satisfies $\bar{x} + y \in X$ and for every $\underline{\delta} \in [\delta, \infty)$ there exists a real number $\bar{\lambda} \in (0,1)$ such that for all $\lambda \in (0, \bar{\lambda})$ the following is true:

$$W(\bar{x} + \lambda \delta y) - \underline{\delta} W(\bar{x} + \lambda y) < (1 - \lambda)(1 - \underline{\delta}) W(\bar{x}) + \lambda [W(\bar{x} + \delta y) - \underline{\delta} W(\bar{x} + y)]. \quad (4.18)$$

PROOF. Assume to the contrary that there exists a sequence of positive real numbers $\lambda_1, \lambda_2, \lambda_3, \ldots$ converging to zero such that for all $i \in \mathbf{Z}^+$ we have

$$W(\bar{x} + \lambda_i \delta y) - \underline{\delta} W(\bar{x} + \lambda_i y) \geq (1 - \lambda_i)(1 - \underline{\delta}) W(\bar{x}) + \lambda_i [W(\bar{x} + \delta y) - \underline{\delta} W(\bar{x} + y)]$$

or, equivalently,

$$\frac{W(\bar{x} + \lambda_i \delta y) - W(\bar{x})}{\lambda_i} - \underline{\delta} \frac{W(\bar{x} + \lambda_i y) - W(\bar{x})}{\lambda_i}$$
$$\geq W(\bar{x} + \delta y) - W(\bar{x}) - \underline{\delta}[W(\bar{x} + y) - W(\bar{x})].$$

For $i \to \infty$ we obtain

$$W'(\bar{x}; \delta y) - \underline{\delta} W'(\bar{x}; y) \geq W(\bar{x} + \delta y) - W(\bar{x}) - \underline{\delta}[W(\bar{x} + y) - W(\bar{x})]$$

where $W'(\bar{x}; z)$ denotes the directional derivative of W at the point $\bar{x} \in X$ in the direction $z \in \mathbf{R}^n$.[8]

We first prove the lemma for $\underline{\delta} = \delta$. In this case, it holds that $W'(\bar{x}; \delta y) = \delta W'(\bar{x}; y)$ because the directional derivatives are positive homogeneous functions. Therefore, we conclude from the above inequality that

$$W(\bar{x} + \delta y) - W(\bar{x}) \leq \delta[W(\bar{x} + y) - W(\bar{x})]$$

must hold which is easily seen to be a contradiction to the strict concavity of W.

Now consider the general case where $\underline{\delta} \geq \delta$. Since (4.18) holds for $\underline{\delta} = \delta$, there exists $\bar{\lambda} > 0$ such that for all $\lambda \in (0, \bar{\lambda})$ we have

$$W(\bar{x} + \lambda \delta y) - W(\bar{x}) - \lambda[W(\bar{x} + \delta y) - W(\bar{x})]$$
$$< \delta \Big\{ W(\bar{x} + \lambda y) - W(\bar{x}) - \lambda[W(\bar{x} + y) - W(\bar{x})] \Big\}.$$

Because of the concavity of W the expression in curly brackets on the right hand side of this inequality is non-negative so that we get

$$W(\bar{x} + \lambda \delta y) - W(\bar{x}) - \lambda[W(\bar{x} + \delta y) - W(\bar{x})]$$
$$< \underline{\delta} \Big\{ W(\bar{x} + \lambda y) - W(\bar{x}) - \lambda[W(\bar{x} + y) - W(\bar{x})] \Big\}$$

whenever $\underline{\delta} \geq \delta$. This inequality is equivalent to (4.18) and the proof of the lemma is therefore complete. $\qquad\square$

Lemma 4.3 is very similar to Proposition 2 in [35]. In that paper, however, the authors claim that the result holds for all $\underline{\delta}$ satisfying $\underline{\delta} \geq |\delta|$ where δ is allowed to be negative and $\bar{x}$ is not required to be in the relative interior of X. Let us, therefore, briefly show that both the restriction $\delta > 0$ and the interiority of $\bar{x}$ are necessary for Lemma 4.3 to be true. To see that interiority of $\bar{x}$ is needed consider $X = [0, 1]$, $\bar{x} = 0$, $y = 1$, $W(x) = \sqrt{x}$, and $0 < \delta < \underline{\delta} < \sqrt{\delta} < 1$. In this case, (4.18) is equivalent to $(\sqrt{\lambda} - \lambda)(\sqrt{\delta} - \underline{\delta}) < 0$ which is wrong for all $\lambda \in (0, 1)$. To see that $\delta > 0$ is necessary consider $X = [-1, 1]$, $\bar{x} = 0$, $y > 0$,

$$W(x) = \begin{cases} -x^2 & \text{for } x \leq 0 \\ -x^3 & \text{for } x \geq 0, \end{cases}$$

---

[8]The directional derivatives of $W'(\bar{x}; y)$ and $W'(\bar{x}; \delta y)$ are both finite, because $\bar{x}$ is a point in the relative interior of X and because $\bar{x} + y \in X$.

$-1 < \delta < 0$, and $1 > \underline{\delta} \geq |\delta|$. Inequality (4.18) turns out to be equivalent to $\delta^2 < \underline{\delta}y(1 + \lambda)$ which is also wrong for all $\lambda \in (0, 1)$ provided that the number $y > 0$ is chosen sufficiently small.

As already indicated by the choice of notation, we interpret the function W in the above lemma as the optimal value function of a strictly concave optimal growth model (T, A) on the state space X. In this regard it should be noted that inequality (4.18) is a translation invariant property. A precise statement of this fact is presented in the following lemma.

Lemma 4.4 *Let (T, A) be an optimal growth model on the state space X and denote by W its optimal value function. Moreover, let $\bar{x}$ and y be given vectors in X such that $\bar{x} + y \in X$. If there exist real numbers λ, δ, and $\underline{\delta}$ such that (4.18) is satisfied and if W' is the optimal value function of a model (T', A') that has been obtained from (T, A) by a translation, then it holds that*

$$W'(\bar{x} + \lambda\delta y) - \underline{\delta}W'(\bar{x} + \lambda y) < (1 - \lambda)(1 - \underline{\delta})W'(\bar{x}) + \lambda[W'(\bar{x} + \delta y) - \underline{\delta}W'(\bar{x} + y)].$$

PROOF. Simply use the identity $W'(x) = W(x) + p \cdot x$ for all $x \in X$ where p is the translation vector. □

We are now ready to prove the second main result of this section. As already mentioned before it provides a partial answer to the question of how sensitive optimal growth paths react to perturbations of their initial states. To this end we consider two different optimal growth paths and try to estimate how fast the distance between the states in corresponding periods can grow as time goes by. Unfortunately, we cannot prove the result for two arbitrary optimal growth paths but have to require that one of them is periodic.

Theorem 4.4 *Let $X \subseteq \mathbf{R}^n$ be a non-empty, closed, and convex set and let two sequences $_0\bar{x} = (\bar{x}_0, \bar{x}_1, \bar{x}_2, \ldots)$ and $_0x = (x_0, x_1, x_2, \ldots)$ of vectors be given such that $\bar{x}_t \in X$ and $x_t \in X$ for all $t \in \mathbf{Z}_0^+$. Assume that $\bar{x}_0 \neq x_0$ and that there exist an integer $\tau \in \mathbf{Z}^+$ and a real number $\delta \in (0, 1)$ such that the following three conditions are satisfied:*

i) $\bar{x}_\tau$ is in the relative interior of X,

ii) $\bar{x}_\tau = \bar{x}_0$, and

iii) $x_0 - \bar{x}_0 = \delta^\tau(x_\tau - \bar{x}_\tau)$.

Define D as the set consisting of the two given paths, i.e., $D = \{_0\bar{x}, _0x\}$. If both sequences $_0\bar{x}$ and $_0x$ are optimal growth paths in the same strictly concave optimal growth model (T, A), then it holds that $\underline{\delta}(D) < \delta$ where $\underline{\delta}(D)$ is the minimal discount factor of A on the set D.

PROOF. We present the proof for the special case $\tau = 1$ only. The general case can be dealt with by a straightforward modification of this proof. Assume the contrary, i.e., $\underline{\delta} = \underline{\delta}(D) \geq \delta$ and denote by W the optimal value function of (T, A) and by y the vector $y = x_1 - \bar{x}_1$. According to Lemma 4.3 we can find a number $\lambda \in (0,1)$ such that

$$W(\bar{x}_1 + \lambda \delta y) - \underline{\delta} W(\bar{x}_1 + \lambda y) < (1 - \lambda)(1 - \underline{\delta})W(\bar{x}_1) + \lambda \left[W(\bar{x}_1 + \delta y) - \underline{\delta} W(x_1)\right] \tag{4.19}$$

holds. Because of Lemmas 4.4 and 3.9 we may assume without loss of generality that W attains its maximum over the state space X at the point $\bar{x}_1 + \lambda y$.[9] Using this property and the fact that $\underline{\delta}$ is the minimal discount factor of (T, A) on D we obtain from (4.19) the following:

$$\begin{aligned}
W(\bar{x}_1 + \lambda \delta y) \\
< (1 - \lambda)\underline{\delta}\left[W(\bar{x}_1 + \lambda y) - W(\bar{x}_1)\right] + \lambda \underline{\delta}\left[W(\bar{x}_1 + \lambda y) - W(x_1)\right] \\
+ (1 - \lambda)W(\bar{x}_1) + \lambda W(\bar{x}_1 + \delta y) \\
\leq (1 - \lambda)\left[A((\bar{x}_0, \bar{x}_1), W(\bar{x}_1 + \lambda y)) - A((\bar{x}_0, \bar{x}_1), W(\bar{x}_1))\right] \\
+ \lambda \left[A((x_0, x_1), W(\bar{x}_1 + \lambda y)) - A((x_0, x_1), W(x_1))\right] \\
+ (1 - \lambda)W(\bar{x}_1) + \lambda W(\bar{x}_1 + \delta y)
\end{aligned}$$

Using the concavity of the aggregator function and the optimal value function as well as the optimality of the paths $_0\bar{x}$ and $_0 x$ and Corollary 3.1, we obtain from the above inequality that

$$\begin{aligned}
W(\bar{x}_1 + \lambda \delta y) < (1 - \lambda)A((\bar{x}_0, \bar{x}_1), W(\bar{x}_1 + \lambda y)) + \lambda A((x_0, x_1), W(\bar{x}_1 + \lambda y)) \\
- (1 - \lambda)W(\bar{x}_0) - \lambda W(x_0) + (1 - \lambda)W(\bar{x}_1) + \lambda W(\bar{x}_1 + \delta y) \\
< A((\bar{x}_0 + \lambda(x_0 - \bar{x}_0), \bar{x}_1 + \lambda y), W(\bar{x}_1 + \lambda y)) \\
- (1 - \lambda)W(\bar{x}_0) - \lambda W(x_0) + (1 - \lambda)W(\bar{x}_1) + \lambda W(\bar{x}_1 + \delta y) \\
\leq W(\bar{x}_0 + \lambda(x_0 - \bar{x}_0)) - (1 - \lambda)W(\bar{x}_0) - \lambda W(x_0) \\
+ (1 - \lambda)W(\bar{x}_1) + \lambda W(\bar{x}_1 + \delta(x_1 - \bar{x}_1)) \\
\leq W(\bar{x}_0 + \lambda(x_0 - \bar{x}_0)).
\end{aligned}$$

In the last step we made use of assumptions ii) and iii) of the theorem. What we have shown, therefore, is that

$$W(\bar{x}_1 + \lambda \delta(x_1 - \bar{x}_1)) < W(\bar{x}_0 + \lambda(x_0 - \bar{x}_0)).$$

Using assumptions ii) and iii) again we can easily see that this is a contradiction and the proof is therefore complete. $\qquad\square$

---

[9]Lemma 3.9 is applicable since $\bar{x}_1 + \lambda y$ is in the relative interior of X for all $\lambda \in (0,1)$.

Let us begin the discussion of the above result with a brief comment on the idea of the proof. To make the argument as simple as possible we consider the additively separable case and assume that $\bar{x}_0$ is a fixpoint ($\tau = 1$ in Theorem 4.4) and that the optimal policy function is linear in an open neighborhood of $\bar{x}_0$. From Theorem 3.1 we know that

$$G(x, h(x)) = V(x, h(x)) + \delta W(h(x)) - W(x) = 0$$

for all $x \in X$. Since h is linear locally at $\bar{x}_0$ it follows that the mapping $x \mapsto V(x, h(x))$ is strictly concave locally at $\bar{x}_0$. Together with the above equation this implies that the mapping $x \mapsto W(x) - \delta W(h(x))$ is strictly concave locally around $\bar{x}_0$. On the other hand, the function $(x, x') \mapsto W(x) - \delta W(x')$ is a saddle function which is strictly convex along the x'-direction. If the slope of $h(x)$ at $\bar{x}_0$ is too steep, then it follows that the linear manifold $x' = h(x)$ is very "close" to the x'-direction, that is, to the convex part of the saddle. In this case we obtain a contradiction to the concavity of the function $x \mapsto W(x) - \delta W(h(x))$.

The interpretation of Theorem 4.4 is rather straightforward: the perturbation of the initial state $\bar{x}_0$ of a stationary or periodic optimal growth path in a certain direction $\Delta x = x_0 - \bar{x}_0$ can never increase in the course of time at a rate exceeding the discount rate. Although the restrictions to periodic paths and to one single direction seem to be awkward, the result is nevertheless interesting and gives rise to various conjectures about a general relationship between the sensitivity of optimal growth paths and the underlying rate of impatience. We shall discuss some of these conjectures later in Section 4.5.

As in the case of Theorem 4.1 we can associate with Theorem 4.4 a certain non-linear mathematical programming problem which gives the best bound on the minimum rate of impatience obtainable from that theorem. This problem is defined as follows:

Minimize δ

subject to $x, \bar{x} \in X$, $\tau \in \mathbf{Z}^+$, $h^\tau(\bar{x}) = \bar{x}$, $\delta^\tau(h^{(\tau)}(x) - \bar{x}) = x - \bar{x} \neq 0$.

Similar to problem P this is a highly non-linear and non-convex optimization problem which will usually have quite a lot of local extrema.

To get some idea whether Theorem 4.4 gives better results than Theorem 4.1 we apply Theorem 4.4 to the same two examples that we have already discussed earlier in this section. The first example is the Hénon map (4.16). Let us apply Theorem 4.4 with $\tau = 1$ and

$$\bar{x}_0 = \bar{x}_1 = (-1.13135, -1.13135)$$

denoting one of the two stationary points of the map. For the slightly perturbed initial state

$$x_0 = (-1.13079, -1.12952)$$

we obtain
$$x_1 = (-1.12952, -1.12539)$$

which yields
$$x_0 - \bar{x}_0 = 0.307(x_1 - \bar{x}_1).$$

Therefore, we conclude that the Hénon map cannot be optimal in a strictly concave optimal growth model with a globally minimal discount factor greater than or equal to 0.307 which corresponds to about 225% discount rate. Here it turns out that Theorem 4.4 yields a better bound than Theorem 4.1.

Now let us consider the second example, namely, the mapping (4.17). The unique fixpoint of this mapping is the point $\bar{x}_0 = (0.20542, 0.20542)$ which is determined by the equations $\bar{x}' = \bar{x}$ and

$$\bar{x} = (\bar{x} + \bar{x}')/10 + 10e^{-10(\bar{x}+\bar{x}')}.$$

To apply Theorem 4.4 with $\tau = 1$ we have to find numbers $\delta \in (0,1)$, $x \in [0,3]$, and $x' \in [0,3]$ such that

$$\begin{pmatrix} x - \bar{x} \\ \\ x' - \bar{x}' \end{pmatrix} = \delta \begin{pmatrix} \dfrac{x + x'}{10} + 10e^{-10(x+x')} - \bar{x} \\ \\ x - \bar{x}' \end{pmatrix}. \qquad (4.20)$$

Substituting the defining equations for $\bar{x}$ and $\bar{x}'$ and rearranging we obtain the equation

$$1 - \delta(1 + \delta)/10 = 0.8\delta \frac{\bar{x}}{x - \bar{x}} \left[e^{-10(x - \bar{x})(1+\delta)} - 1 \right].$$

The right hand side of this equation is always negative whereas the left hand side is positive for all $\delta \in (0,1)$. This shows that there is no solution to Equation (4.20) and Theorem 4.4 with $\tau = 1$ does not apply. This example demonstrates that there are situations where Theorem 4.1 gives a useful bound on the minimum rate of impatience whereas Theorem 4.4 does not.[10]

4.2 The One-Dimensional Case

This section serves as an extended example for the application of the Minimum Impatience Theorems. More specifically, we shall present various corollaries to Theorems 4.1 - 4.4 which deal with the case of an economy in which there is only a single capital good, i.e., the case $n = 1$.

To begin with let us consider once more the mathematical programming problem P discussed in the preceding section. It consists of minimizing δ over

---

[10]Of course, one could try to apply Theorem 4.4 with $\tau > 1$ to obtain a useful bound. This exercise is left to the reader.

the set of all $\delta \in (0,1]$, $x \in X$, and $\mu_t \in \mathbf{R}$, $t \in \mathbf{Z}^+$, such that the following conditions are satisfied:

$$x \neq h(x), \tag{4.21}$$

$$0 \leq \mu_{t+1} \leq \mu_t \text{ for all } t \in \mathbf{Z}^+, \tag{4.22}$$

$$\sum_{t=1}^{\infty} \delta^t \mu_t = 1, \tag{4.23}$$

$$\sum_{t=1}^{\infty} \delta^t \mu_t h^{(t)}(x) = x. \tag{4.24}$$

We have already mentioned in Section 4.1 that this problem is very hard to solve exactly. One of the difficulties is the high dimensionality of P which is caused by infinitely many variables μ_t. We can reduce the dimension by introducing additional constraints. For example, let us assume that the inequality constraint $\mu_{t+1} \leq \mu_t$ is replaced by the corresponding equality constraint $\mu_{t+1} = \mu_t$ for all $t \in \mathbf{Z}^+$. This implies that all weights μ_t have to be equal, say, $\mu_t = \mu$ for all $t \in \mathbf{Z}^+$. Hence there are only three variables left: x, δ, and μ. Condition (4.22) is obviously satisfied in this case as long as $\mu \geq 0$. Condition (4.23), on the other hand, requires that $\mu = (1-\delta)/\delta$. Substituting this into (4.24), we obtain

$$\sum_{t=0}^{\infty} \delta^t \left[h^{(t+1)}(x) - x \right] = 0. \tag{4.25}$$

Whenever x is a fixpoint of h, this equation is automatically satisfied. Fixpoints of h, however, are excluded by condition (4.21). Therefore, we obtain the following result as a corollary to Theorem 4.2.

Lemma 4.5 *Let $X \subseteq \mathbf{R}^1$ be a non-empty and closed interval and let $h : X \mapsto X$ be an arbitrary function. Assume that for some $x \in X$ with $x \neq h(x)$ there exists a real number $\delta \in (0,1]$ satisfying Equation (4.25). Then there does not exist any strictly concave optimal growth model (T, A) with state space X such that h is the optimal policy function of this model, and such that the minimal discount factor of A satisfies $\underline{\delta}(D) \geq \delta$ with D as defined in Theorem 4.2.*

The restriction to the one-dimensional case is, strictly speaking, not necessary. However, except for hairline cases the conditions of Lemma 4.5 will not be satisfied when $n > 1$. This follows from the fact that (4.25) represents a system of n equations in the single variable δ. In general, such a system does not have a solution if $n > 1$.

Lemma 4.5 is extremely powerful and easy to apply, especially if the point x is eventually mapped onto a fixpoint or a periodic point of the mapping h. To see this let us assume that $h^{(k+\tau)}(x) = h^{(k)}(x)$ for some $k \in \mathbf{Z}_0^+$ and $\tau \in \mathbf{Z}^+$.

Then we know that $h^{(t+\tau)}(x) = h^{(t)}(x)$ for all $t \geq k$. Using this fact we can transform Equation (4.25) as follows:[11]

$$0 = \sum_{t=0}^{\infty} \delta^t (x_{t+1} - x_0)$$

$$= \sum_{t=0}^{k-1} \delta^t (x_{t+1} - x_0) + \sum_{i=0}^{\infty} \sum_{t=0}^{\tau-1} \delta^{k+i\tau+t} (x_{k+t+1} - x_0)$$

$$= \sum_{t=0}^{k-1} \delta^t (x_{t+1} - x_0) + \sum_{t=0}^{\tau-1} \frac{\delta^{k+t} (x_{k+t+1} - x_0)}{1 - \delta^\tau}$$

Multiplying this equation by $1 - \delta^\tau$ and rearranging terms yields

$$\sum_{t=0}^{\tau-1} \delta^t (x_{t+1} - x_0) + \sum_{t=\tau}^{k+\tau-2} \delta^t (x_{t+1} - x_{t+1-\tau}) = 0. \qquad (4.26)$$

This is a polynomial equation of degree $k + \tau - 2$ in the unknown variable δ. As a first example of its application let us consider the logistic map $h(x) = 4x(1 - x)$ which has already been discussed in Section 3.3. Consider the trajectory emanating from the initial state $x_0 = [1 + \cos(3\pi/8)]/2$. It is easy to verify that x_0 is eventually mapped on the fixpoint 0 and that its orbit is given by

$$x_0 = [1 + \cos(3\pi/8)]/2,$$
$$x_1 = [1 + \cos(\pi/4)]/2,$$
$$x_2 = 1/2,$$
$$x_3 = 1,$$
$$x_t = 0 \qquad \text{for all } t \geq 4.$$

Therefore, we have $\tau = 1$ and $k = 4$. Substituting these values into (4.26) we obtain the cubic equation

$$\delta^3 - \frac{\delta^2}{2} + \frac{\delta}{2} \cos \frac{\pi}{4} - \frac{1}{2} \left(\cos \frac{\pi}{4} - \cos \frac{3\pi}{8} \right) = 0$$

which has a solution at approximately 0.475. This shows that the logistic map cannot be the optimal policy function of a strictly concave optimal growth problem with a discount factor greater than or equal to 0.475 or, equivalently, a time preference rate smaller than or equal to roughly 110%. From our numerical experiments (see Figures 4.1 – 4.4 in Section 4.3) it follows that this bound is about the best we can get from the application of Theorem 4.2 to the

---

[11]We use the short hand notation x_t instead of $h^{(t)}(x)$.

logistic map. In other words, $\delta = 0.475$ is the optimal value of problem P. It will be shown shortly, however, that a much better bound can be obtained by considering higher iterates of the logistic function.

Our next example is the tent map $h(x) = 1 - |2x - 1|$ for which we know from Lemma 3.12 that it can be optimal in a strictly concave and additively separable optimal growth model at least as long as the discount factor is smaller than $1/4$. We are now going to derive a bound for this policy function by combining Theorem 4.3 and Lemma 4.5. To this end, we apply Theorem 4.3 with $\tau = 3$, i.e., we consider the third iterate $h^{(3)}$ of the tent map. For the initial state $x_0 = 25/32$ we obtain

$$x_0 = 25/32,$$
$$x_1 = h^{(3)}(x_0) = 7/8,$$
$$x_2 = h^{(6)}(x_0) = 1/2,$$
$$x_t = h^{(3t)}(x_0) = 0 \qquad \text{for all } t \geq 3.$$

We see that x_0 is mapped onto the fixpoint 0 after three iterations of $h^{(3)}$ so that we have to choose $k = 3$ and $\tau = 1$ in Equation (4.26). This yields the quadratic equation

$$\delta^2 + \frac{3}{4}\delta - \frac{3}{16} = 0$$

which has the solution $\delta = (\sqrt{21} - 3)/8$. This shows that $x_0 = 25/32$ is a non-trivial fixpoint of the average correspondence associated with $h^{(3)}$ at $\delta = (\sqrt{21} - 3)/8$. From Theorem 4.3 we can therefore conclude that the tent map itself cannot be optimal in a strictly concave optimal growth model (T, A) with

$$\underline{\delta}(\mathbf{F}) \geq \sqrt[3]{\frac{\sqrt{21} - 3}{8}} \approx 0.445.$$

In other words, the tent map requires a minimum rate of impatience of approximately 125%.

Another useful result for one-dimensional maps which is based on both Lemma 4.5 and Theorem 4.4 is the following one.

Lemma 4.6 *Let $X \subseteq \mathbf{R}^1$ be a non-empty and closed interval and let $h : X \mapsto X$ be a given continuous function. Assume that there exist feasible states $a \in X$ and $b \in X$ such that $a < b$, $h(a) = h(b) = a$, and $h([a, b]) \supseteq [a, b]$. Define the numbers*

$$c^+ = \max\{x \in [a, b] \,|\, h(x) = b\},$$
$$c^- = \min\{x \in [a, b] \,|\, h(x) = b\},$$
$$\delta^+ = (b - c^+)/(b - a) < 1,$$
$$\delta^- = (c^- - a)/(b - a) < 1.$$

Then there does not exist any strictly concave optimal growth model (T, A) with state space X such that h is the optimal policy function of this model and such that the minimal discount factor of A on the set $D^+ = \{(c^+, b, a, a, \ldots)\}$ satisfies $\underline{\delta}(D^+) \geq \delta^+$. If the point a is not an endpoint of the interval X, then we have the additional result that h is certainly not the optimal policy function of any strictly concave optimal growth model (T, A) with $\underline{\delta}(D^-) \geq \delta^-$ where $D^- = \{(c^-, b, a, a, \ldots)\}$.

PROOF. Let c^+ (c^-) be the largest (smallest) number c in the interval $[a, b]$ which satisfies $h(c) = b$. Because h is continuous, it must hold that $a < c^- \leq c^+ < b$ and, consequently, $\delta^+ < 1$ and $\delta^- < 1$.

After two iterations of h, the point c^+ is mapped on the fixpoint a. Therefore, the sequence $_0x = (x_0, x_1, x_2, \ldots)$ defined by $x_0 = c^+$, $x_1 = b$, and $x_t = a$ for all $t \geq 2$ is a trajectory generated by the policy function h. Substituting this trajectory into (4.26) with $k = 2$ and $\tau = 1$, we obtain the linear equation $b - c^+ + \delta(a - b) = 0$ which has the unique solution $\delta = \delta^+$. Therefore, the first assertion of the lemma follows from Lemma 4.5.

Now consider the sequence $_0x = (x_0, x_1, x_2, \ldots)$ defined by $x_0 = c^-$, $x_1 = b$, and $x_t = a$ for all $t \geq 2$. This is also a trajectory generated by h. If the fixpoint a is in the relative interior of X (i.e., if it is not an endpoint of the interval X), then we can apply Theorem 4.4 with $\bar{x}_t = a$ for all $t \in \mathbf{Z}_0^+$ and $\tau = 1$. Because $x_0 - \bar{x}_0 = \delta^-(x_1 - \bar{x}_1)$ this yields the second result stated in the lemma. □

The situation described by the assumptions of Lemma 4.6 includes the case of "one-humped" mappings on the interval $X = [a, b]$ for which it is relatively easy to prove the occurrence of complicated dynamics (see, e.g., [27, Proposition 6]). In particular, the logistic map and the tent map satisfy these assumptions with $a = 0$, $b = 1$, and $c^+ = 1/2$. Therefore, neither of these maps can be optimal in a strictly concave optimal growth model with a globally minimal discount factor greater than or equal to $1/2$ which corresponds to a discount rate of exactly 100%. Of course, we have already obtained better bounds for these maps but the advantage of Lemma 4.6 is its simplicity. As another example consider the one-parameter family of maps $h_c : [0, 1] \mapsto [0, 1]$ defined for all $c \in (0, 1)$ as follows:

$$
h_c(x) = \begin{cases} x/c & \text{for } x \in [0, c] \\ \dfrac{1 - x}{1 - c} & \text{for } x \in [c, 1] \end{cases} \tag{4.27}
$$

It is obvious that $h_{0.5}$ is exactly the tent map and that h_c represents an asymmetric or "lop-sided tent" when $c \neq 1/2$. Furthermore, the maps h_c, $c \in (0, 1)$, satisfy the conditions of the first part of Lemma 4.6 with $a = 0$, $b = 1$, and $\delta^+ = 1 - c$. This shows that we get a very small bound on the minimal discount factor if c is close to one and that we get a very poor bound if c is close to zero. Note that we cannot apply the second part of Lemma 4.6 since $a = 0$ is an endpoint of the state space $[0, 1]$. We shall further discuss this example in Section 4.4.

Now let us consider the iterates of a one-humped surjective function $h :$ $[a, b] \mapsto [a, b]$. The second iterate $h^{(2)}$ is a two-humped function, the third iterate $h^{(3)}$ is a four-humped one, and the τ-th iterate $h^{(\tau)}$ is a mapping with $2^{\tau-1}$ humps. The distance from the rightmost hump to the right hand endpoint, b, of the state space $X = [a, b]$ converges to zero as τ approaches infinity. Let us denote by $c(\tau)$ the x-value at which the rightmost hump of $h^{(\tau)}$ occurs. Because of $\lim_{\tau \to \infty} c(\tau) = b$ we see that the bound of Lemma 4.6,

$$\underline{\delta}(D^+) < (b - c(\tau))/(b - a),$$

becomes arbitrary small as τ approaches infinity. Note, however, that this is a bound for the τ-th iterate of h and not for the original mapping h itself. To put it differently, the minimum rate of impatience required for the optimality of the τ-th iterate of h can be made arbitrary high by choosing τ sufficiently large. But what does this imply for the minimum rate of impatience required for the optimality of the first iterate h? According to Lemma 3.10 (see also Theorem 4.3) we obtain the bound

$$\underline{\delta}(\mathbf{F}) \leq \inf_{\tau \in \mathbf{Z}^+} \sqrt[\tau]{\frac{b - c(\tau)}{b - a}}. \tag{4.28}$$

This can be quite a good bound as will be illustrated by means of the logistic map $h(x) = 4x(1 - x)$. For this function we have $a = 0$ and $b = 1$. Moreover, the largest number $x \in [0, 1]$ satisfying $h^{(\tau)}(x) = 1$ is given by $x = c(\tau) = [1 + \cos(\pi/2^\tau)]/2$. According to the above argument we can conclude that for any strictly concave optimal growth model (T, A) which has the logistic map as its optimal policy function it must hold that

$$\underline{\delta}(\mathbf{F}) \leq \inf_{\tau \in \mathbf{Z}^+} \sqrt[\tau]{\frac{1 - \cos(\pi/2^\tau)}{2}}.$$

The τ-th root on the right hand side of this inequality considered as a function of τ is strictly decreasing. Moreover, application of de L'Hôpital's rule shows that its limit as τ approaches infinity is equal to $1/4$. Therefore, we have proved that $\underline{\delta}(\mathbf{F}) \leq 1/4$ for any strictly concave optimal growth model that is solved by the logistic map.

In the case of the logistic map the bound on the minimal discount factor obtained by applying Lemma 4.6 to the τ-th iterate is a decreasing function of τ so that we obtained the best value in the limit as τ approached infinity. This need not always be the case. The reader is encouraged to verify that for the tent map $h(x) = 1 - |2x - 1|$ it holds that

$$\sqrt[\tau]{\frac{b - c(\tau)}{b - a}} = 1/2$$

for all $\tau \in \mathbf{Z}^+$ so that the estimate given in (4.28) is not better than the one derived from application of Lemma 4.6 to the first iterate of h. In particular, it is not possible to obtain a better bound for the tent map from (4.28) than the one presented earlier.

Now let us see whether we can obtain the bound $\underline{\delta}(\mathbf{F}) \leq 1/4$ for the logistic map also from Theorem 4.4. If the fixpoint $\bar{x}_0 = 0$ were in the relative interior of $[0, 1]$, then it would be easy as we can find for every $\delta < 1/4$ a number $x_0(\delta) > 0$ such that the path $(x_0, x_1, x_2, \ldots)$ emanating from $x_0 = x_0(\delta)$ satisfies $x_1 = 4x_0(1 - x_0) = \delta^{-1} x_0$ which would yield the desired result. Since 0 is not in the relative interior, however, we have to be a little bit more careful. Let us specify $\bar{x}_0(\tau) = [1 - \cos(2\pi/(2^\tau - 1))]/2$ and $x_0(\tau) = [1 - \cos(3\pi/2^\tau)]/2$. It is straightforward to verify that $\bar{x}_0(\tau)$ is a τ-periodic point of the logistic map and that the τ-th iterate of $x_0(\tau)$ satisfies $x_\tau(\tau) = 1$. Since $\bar{x}_0(\tau)$ is in the relative interior of the interval $[0, 1]$ we can apply Theorem 4.4 to obtain

$$\underline{\delta}(\mathbf{F}) \leq \inf_{\tau \in \mathbf{Z}^+} \sqrt[\tau]{\frac{x_0(\tau) - \bar{x}_0(\tau)}{1 - \bar{x}_0(\tau)}}.$$

Substituting the expressions for $\bar{x}_0(\tau)$ and $x_0(\tau)$ into the right hand side of this inequality one can see that the infimum is approached in the limit as τ tends to infinity and that it is equal to $1/4$. Hence, we have derived from Theorem 4.4 the same bound for the logistic map as from Theorem 4.1.

It is well known that the existence of a periodic point with minimal period three has far reaching implications for the dynamic behavior of one-dimensional maps. This is a consequence of the celebrated theorem of Sarkovskii (see, e.g., [28, 58]). Moreover, because of the famous paper "Period Three Implies Chaos" [42], the existence of a three-periodic point has sometimes been taken as a possible definition of chaotic behavior. The following lemma shows what we can infer from the Minimum Impatience Theorems in this case.

Lemma 4.7 *Let $X \subseteq \mathbf{R}^1$ be a non-empty and closed interval and let $h : X \mapsto X$ be a given function. Assume that there exists a feasible state $a \in X$ such that $h^{(3)}(a) = a$ and define $b = h(a)$ and $c = h(b)$. If $0 < (a - b)/(c - a) < 1$, then there does not exist any strictly concave optimal growth model (T, A) with state space X such that h is the optimal policy function of this model and such that the minimal discount factor of A along the periodic path $D = \{(a, b, c, a, \ldots)\}$ satisfies $\underline{\delta}(D) \geq (a - b)/(c - a)$.*

PROOF. It is straightforward to verify that $\delta = (a - b)/(c - a)$ is a solution of Equation (4.26) when $k = 0$, $\tau = 3$, $x_1 = a$, $x_2 = b$, and $x_3 = c$. Therefore, the result follows from Lemma 4.5. □

As for our standard example, the logistic map, there are six points with a minimal period three. Only one of them, namely, $a = [1 + \cos(3\pi/7)]/2$ satisfies the condition $0 < (a-b)/(c-a) < 1$ required in Lemma 4.7. The corresponding bound on the minimal discount factor turns out to be $\delta = 0.802$ which is not as good as those derived earlier.

It is clear that there are policy functions which resist all methods described so far to obtain a useful bound $\delta < 1$. Because of this reason it is interesting to develop sufficient conditions for the existence of such a bound. One possible way of doing this will now be demonstrated. Assume that h is a continuous function on a compact state space $X = [a, b]$ and denote the left hand side of Equation (4.25) by $f(\delta)$, that is,

$$f(\delta) = \sum_{t=0}^{\infty} \delta^t \left[h^{(t+1)}(x) - x \right]. \tag{4.29}$$

The function $f : [0, 1) \mapsto X$ is continuous and satisfies $f(0) = h(x) - x$ and

$$\operatorname{sgn} \lim_{\delta \nearrow 1} f(\delta) = \operatorname{sgn} \left\{ -x + \lim_{s \to \infty} \frac{1}{s+1} \sum_{t=0}^{s} h^{(t+1)}(x) \right\}.$$

Because of the Ergodic Theorem the latter limit exists ρ-almost everywhere provided that ρ is an invariant ergodic measure for h (see, e.g., [29]). Recalling from Lemma 4.5 that any solution of $f(\delta) = 0$ with $\delta \in (0, 1)$ qualifies as a non-trivial upper bound on the minimal discount factor, we obtain from a continuity argument that each of the following two conditions is sufficient for the existence of such a non-trivial bound δ:

i) There exists a vector $x \in X$ such that

$$h(x) > x > \lim_{s \to \infty} \frac{1}{s+1} \sum_{t=0}^{s} h^{(t+1)}(x).$$

ii) There exists a vector $x \in X$ such that

$$h(x) < x < \lim_{s \to \infty} \frac{1}{s+1} \sum_{t=0}^{s} h^{(t+1)}(x).$$

The limit on the right hand sides of the above formulas denotes the long-run average capital stock if one starts in the initial state x and uses the policy function h. The interpretation of the conditions is therefore as follows: a policy function h which requires that, for some capital stock $x \in [a, b]$, one has to choose the capital stock for the next period, $h(x)$, in "the opposite direction of the long-run average capital stock" cannot be optimal for arbitrary small discount rates.

It is also possible to use our results to demonstrate that there are continuous functions $h : X \mapsto X$ which cannot be optimal in any strictly concave optimal growth model however small the globally minimal discount factor $\underline{\delta}(\mathbf{F})$ is chosen. This fact has already been pointed out in [35]. Consider for example the function $h(x) = \sqrt[3]{x}$ defined on the state space $X = [-1, 1]$. This function is infinitely steep at its fixpoint $\bar{x}_0 = 0$ and one can readily verify that the assumptions of Theorem 4.4 can be satisfied for all $\delta > 0$ by choosing $x_0 > 0$ sufficiently small. This shows that $h(x) = \sqrt[3]{x}$ cannot be the optimal policy function in any strictly concave optimal growth model (T, A).

We conclude this section by pointing out that some of the results presented here can be generalized to higher dimensional state spaces. There are, however, quite a lot of different ways of doing this and it is not at all clear which one of them yields the most powerful results or the results which are most easily applicable.

4.3 The Average Correspondence

The average correspondence introduced in Definition 4.1 bears the entire information which is necessary to apply the Minimum Impatience Theorems 4.1 and 4.2. It seems therefore worthwhile to study its properties even if they are not of immediate relevance to our further analysis. In the present section we summarize several basic results about the average correspondence $H(x, \delta)$ associated with a given policy function $h : X \mapsto X$. It will be assumed throughout this section that $X \subseteq \mathbf{R}^n$ is a given state space and that h is a continuous function.

Our first aim is to derive an alternative representation of $H(x, \delta)$ which will turn out to be very useful in Section 4.4. To this end let us define for every $x \in X$ and every $m \in \mathbf{Z}^+$

$$g_m(x, \delta) = \begin{cases} \dfrac{1 - \delta}{\delta(1 - \delta^m)} \displaystyle\sum_{t=1}^{m} \delta^t h^{(t)}(x) \,, & \text{for } \delta \in (0, 1) \\[2em] \dfrac{1}{m} \displaystyle\sum_{t=1}^{m} h^{(t)}(x) \,, & \text{for } \delta = 1 \end{cases} \tag{4.30}$$

and

$$H_m(x, \delta) = \text{conv}\Big\{ g_t(x, \delta) \,|\, 1 \le t \le m \Big\}.$$

Here, conv C denotes the convex hull of the set C, i.e., the smallest convex set containing C. We have the following result.

Lemma 4.8 *For every $m \in \mathbf{Z}^+$, $x \in X$, and $\delta \in (0,1]$ it holds that*

$$\{h(x)\} = H_1(x,\delta) \subseteq H_m(x,\delta) \subseteq H_{m+1}(x,\delta) \subseteq H(x,\delta).$$

PROOF. Except for the last inclusion the results follow immediately from the definition of $H_m(x,\delta)$. It remains to be proved that $H_m(x,\delta) \subseteq H(x,\delta)$ for all $m \in \mathbf{Z}^+$. To this end we first observe that $H(x,\delta)$ is a convex set. Furthermore, it is easy to see that $g_m(x,\delta) \in H(x,\delta)$ for all $m \in \mathbf{Z}^+$. These two observations together with the definition of $H_m(x,\delta)$ prove the lemma. □

The reason for introducing the correspondences H_m is that they are easier to handle than the average correspondence H. In particular, since $H_m(x,\delta)$ is defined as the convex hull of finitely many points $g_m(x,\delta)$, and since the functions $g_m : X \times (0,1] \mapsto X$ are continuous for all $m \in \mathbf{Z}^+$, it follows that the correspondences $H_m : X \times (0,1] \mapsto 2^X$ are compact-valued and continuous for all $m \in \mathbf{Z}^+$ (see, e.g., [36, Chapter B.III]). The average correspondence H, on the other hand, is not compact-valued without additional assumptions. However, it is possible to approximate the average correspondence H by the sequence H_1, H_2, H_3, ... on a suitable subset of its domain $X \times (0,1]$. This approximation theorem can then be used to show that the average correspondence is continuous and compact-valued on the same restricted domain.

Before we present the approximation theorem, we would like to point out that restricting the average correspondence H to a subset of its domain does not restrict the use of H in the Minimum Impatience Theorems. To explain this issue, we consider the given policy function $h : X \mapsto X$. If this function does not satisfy the condition

$$\inf_{\alpha > 0} \sup \left\{ |h(x)|/(\alpha + |x|) \,\Big|\, x \in X \right\} < \infty,$$

then it follows from condition iii) in Definition 2.1 that h is not the optimal policy function in any optimal growth model simply because h does not generate feasible growth paths. In this case there is no need to appeal to Theorem 4.2. Let us, therefore, define β_h as the infimum of all real numbers $\beta > 1$ for which

$$\inf_{\alpha > 0} \sup \left\{ |h(x)|/(\alpha + |x|) \,\Big|\, x \in X \right\} \le \beta.$$

Again referring to condition iii) of Definition 2.1 we see that h cannot be the optimal policy function in an optimal growth model (T, A) with a β-bounded technology provided that $\beta < \beta_h$. The relevant domain is therefore $\beta \ge \beta_h$ and on this domain we obtain from condition iii) of Definition 2.3 that $\bar{\delta} < 1/\beta_h$ must hold. To summarize, we can say that for a given policy function $h : X \mapsto X$ we need to consider the average correspondence H only on the set $X \times (0, \beta_h^{-1})$ since for all values $\delta \geq 1/\beta_h$ the optimality of h can be excluded without the Minimum Impatience Theorems. We are now in a position to state the aforementioned approximation result.

Theorem 4.5 *For every $x \in X$ and every $\delta \in (0, \beta_h^{-1})$ it holds that*

$$H(x, \delta) = \overline{\bigcup_{m \in \mathbf{Z}^+} H_m(x, \delta)}, \tag{4.31}$$

where a bar over a set denotes its closure. Moreover, the restriction of the average correspondence H to the domain $X \times (0, \beta_h^{-1})$ is a continuous and convex-valued correspondence.

PROOF. It follows from Lemma 4.8 that the inclusion

$$\bigcup_{m \in \mathbf{Z}^+} H_m(x, \delta) \subseteq H(x, \delta)$$

holds for all $x \in X$ and for all $\delta \in (0, 1]$. To prove (4.31) it is therefore sufficient to show that $H(x, \delta)$ is closed and that

$$H(x, \delta) \subseteq \overline{\bigcup_{m \in \mathbf{Z}^+} H_m(x, \delta)}. \tag{4.32}$$

We first show that $H(x, \delta)$ is closed. To this end we consider a sequence of vectors $y^{(1)}, y^{(2)}, y^{(3)}, \ldots$ such that $y^{(k)} \in H(x, \delta)$ for all $k \in \mathbf{Z}^+$ and $\lim_{k \to \infty} y^{(k)} = y$. According to Definition 4.1 every $y^{(k)}$ has a representation

$$y^{(k)} = \sum_{t=1}^{\infty} \mu_t^{(k)} \delta^t h^{(t)}(x), \tag{4.33}$$

where

$$\sum_{t=1}^{\infty} \mu_t^{(k)} \delta^t = 1. \tag{4.34}$$

From conditions (4.1) and (4.2) it follows that $0 \leq \mu_t^{(k)} \leq \mu_1^{(k)} \leq 1/\delta$ holds for all t and all k. Therefore, it is possible to find a subsequence $k_1, k_2, k_3,$ $\ldots$ such that the limit as l approaches infinity of $\mu_t^{(k_l)}$ exist for all $t \in \mathbf{Z}^+$. Denoting this limit by μ_t, it is clear that the sequence $\mu_1, \mu_2, \mu_3, \ldots$ satisfies condition (4.1). Moreover, because of $\delta < 1/\beta_h$ it follows that the infinite series in (4.33) and (4.34) converge absolutely. When we pass over to the limit $l \to \infty$ in these equations (with $\mu_t^{(k)}$ and $y^{(k)}$ replaced by $\mu_t^{(k_l)}$ and $y^{(k_l)}$, respectively) we can therefore interchange summation and limit to obtain (4.15) and (4.2), respectively. This shows that $y \in H(x, \delta)$ and, consequently, that $H(x, \delta)$ is closed.

To prove (4.32) let $y \in H(x, \delta)$ be given. We are going to show that for every $\epsilon > 0$ one can find $M(\epsilon) \in \mathbf{Z}^+$ and $\tilde{y}(\epsilon) \in H_{M(\epsilon)}(x, \delta)$ with $|\tilde{y}(\epsilon) - y| \le \epsilon$. According to Definition 4.1 there exist real numbers μ_1, μ_2, μ_3, ... such that conditions (4.1), (4.2), and (4.3) are satisfied. Moreover, because of $\delta < 1/\beta_h$ it follows that the series in (4.2) and (4.3) converge absolutely. Hence, for every $\epsilon > 0$ there exists a number $M(\epsilon) \in \mathbf{Z}^+$ such that

$$\left| \left(\sum_{t=1}^{M(\epsilon)} \mu_t \delta^t \right)^{-1} - 1 \right| < \frac{\epsilon}{2|y| + \epsilon}$$

and such that

$$\left| \sum_{t=1}^{M(\epsilon)} \mu_t \delta^t h^{(t)}(x) - y \right| < \frac{\epsilon}{2}.$$

Defining $y(\epsilon)$ by

$$y(\epsilon) = \sum_{t=1}^{M(\epsilon)} \mu_t \delta^t h^{(t)}(x)$$

we obtain $|y(\epsilon)| \le |y| + \epsilon/2$ and, therefore,

$$\left| y(\epsilon) \left(\sum_{t=1}^{M(\epsilon)} \mu_t \delta^t \right)^{-1} - y \right| \le |y(\epsilon)| \frac{\epsilon}{2|y| + \epsilon} + |y(\epsilon) - y| \le \frac{\epsilon}{2} + \frac{\epsilon}{2} = \epsilon.$$

It remains to be shown that the vector $\tilde{y}(\epsilon)$ defined by

$$\tilde{y}(\epsilon) = y(\epsilon) \left(\sum_{t=1}^{M(\epsilon)} \mu_t \delta^t \right)^{-1}$$

is an element of $H_{M(\epsilon)}(x, \delta)$. By partial summation we obtain the following representation of $y(\epsilon)$:

$$\begin{aligned}
y(\epsilon) &= \sum_{t=1}^{M(\epsilon)} \mu_t \delta^t h^{(t)}(x) \\
&= \left[\sum_{t=1}^{M(\epsilon)-1} (\mu_t - \mu_{t+1}) \sum_{s=1}^{t} \delta^s h^{(s)}(x) \right] + \mu_{M(\epsilon)} \sum_{t=1}^{M(\epsilon)} \delta^t h^{(t)}(x) \\
&= \sum_{t=1}^{M(\epsilon)-1} \delta \frac{1 - \delta^t}{1 - \delta} (\mu_t - \mu_{t+1}) g_t(x, \delta) + \delta \frac{1 - \delta^{M(\epsilon)}}{1 - \delta} \mu_{M(\epsilon)} g_{M(\epsilon)}(x, \delta)
\end{aligned}$$

Because of $\mu_t \geq \mu_{t+1}$ for all $t \in \mathbf{Z}^+$ and because of

$$\sum_{t=1}^{M(\epsilon)-1} \delta \frac{1-\delta^t}{1-\delta}(\mu_t - \mu_{t+1}) + \delta \frac{1-\delta^{M(\epsilon)}}{1-\delta} \mu_{M(\epsilon)} = \sum_{t=1}^{M(\epsilon)} \mu_t \delta^t,$$

we finally see that $\tilde{y}(\epsilon)$ is a convex combination of the points $g_1(x,\delta)$, $g_2(x,\delta)$, $\ldots, g_{M(\epsilon)}(x,\delta)$. It follows therefore from the definition of $H_{M(\epsilon)}(x,\delta)$ that $\tilde{y}(\epsilon) \in H_{M(\epsilon)}(x,\delta)$ and the proof of (4.32) is complete.

Finally, the continuity of $H : X \times (0, \beta_h^{-1}) \mapsto 2^X$ follows directly from the approximation (4.31) and from the definition of continuity of a correspondence (see [36, Chapter B.III]). This completes the proof. □

It should be noted that the restriction to $X \times (0, \beta_h^{-1})$ is necessary for Theorem 4.5 to hold. To see this consider the logistic map $h(x) = 4x(1-x)$ defined on $X = [0,1]$. For this map we have $\beta_h = 1$ since the state space is compact. The values for δ which are admissible in Theorem 4.5 are therefore $\delta \in (0,1)$. Now choose the discount factor $\delta = 1$ and the initial state $x = 1/2$. Then it holds that $h(x) = 1$ and $h^{(t)}(x) = 0$ for all $t \geq 2$. It is easy to see that $0 \notin H(x,1)$ but that $0 \in \overline{H(x,1)}$ which proves that $H(x,1)$ is not closed. Hence, the approximation formula (4.31) cannot be true for $\delta = 1$.

The next lemma presents some properties of the average correspondence which follow easily from its definition or from Theorem 4.5.

Lemma 4.9 *Let $X \subseteq \mathbf{R}^n$ be a state space and let $h : X \mapsto X$ be a given continuous function with associated average correspondence $H : X \times (0,1]$. Then the following properties are true:*

i) $H(x,\delta) \subseteq H(x,\delta')$ for all $0 < \delta \leq \delta' \leq 1$.

ii) $\lim_{\delta \searrow 0} H(x,\delta) = \{h(x)\}$.

iii) $H(x,\delta) = \{x\}$ holds if and only if x is a fixpoint of h.

iv) $H(x,\delta) = \{h(x)\}$ holds if and only if $h(x)$ is a fixpoint of h.

v) $H(x,\delta)$ is a singleton if and only if either x or $h(x)$ is a fixpoint of h.

PROOF. Statement i) was proved in Lemma 4.1. Statement ii) follows from Theorem 4.5 and from the simple observation that $\lim_{\delta \searrow 0} H_m(x,\delta) = \{h(x)\}$. Statements iii), iv), and v) are immediate implications of Definition 4.1. □

We conclude this section by illustrating some of the results in the case that h is the logistic map. In Figures 4.1 - 4.4 we have depicted the correspondence $H_m(x,\delta)$ for $m = 20$ and the δ-values 1.0, 0.7, 0.5, and 0.475, respectively. The shaded area is the graph of the multifunction $x \mapsto H_m(x,\delta)$. The upward

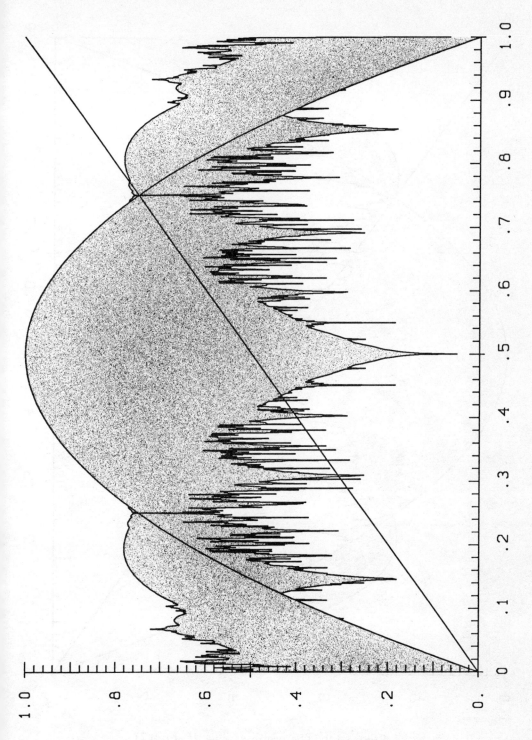

Figure 4.1: The graph of $x \mapsto H_{20}(x, 1.0)$

86

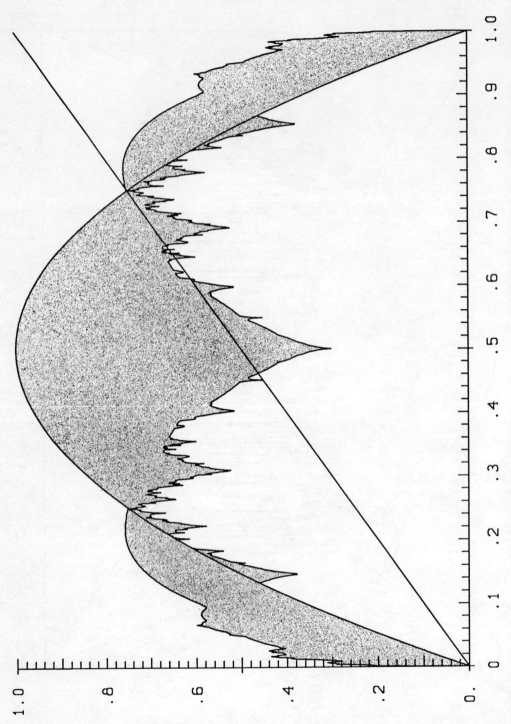

Figure 4.2: The graph of $x \mapsto H_{20}(x, 0.7)$

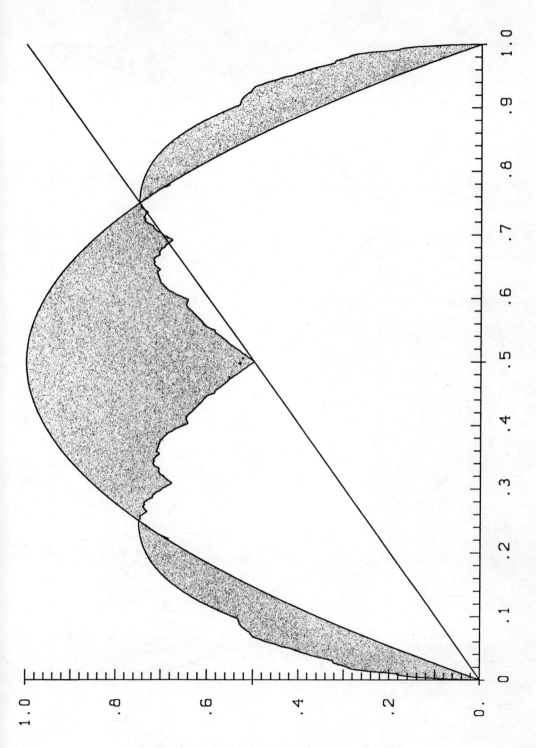

Figure 4.3: The graph of $x \mapsto H_{20}(x, 0.5)$

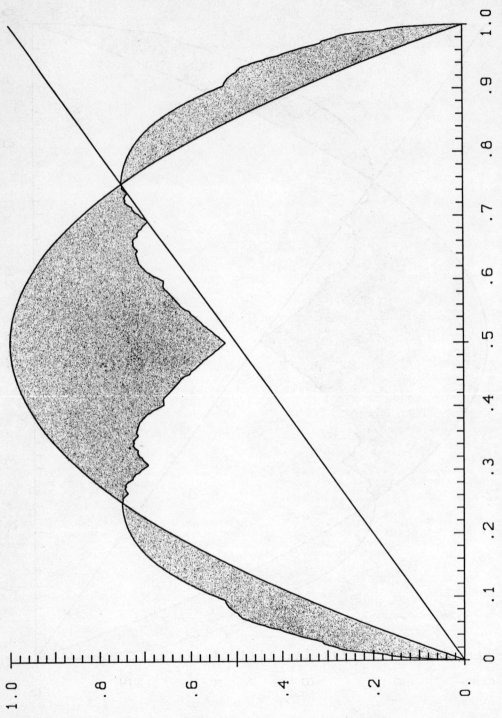

Figure 4.4: The graph of $x \mapsto H_{20}(x, 0.475)$

pointing parabola is the graph of h. Because the pictures do not change substantially if we increase m beyond the value 20, we can consider the figures as good approximations of the correspondences $x \mapsto H(x,1)$, $x \mapsto H(x,0.7)$, $x \mapsto H(x,0.5)$, and $x \mapsto H(x,0.475)$, respectively. We can see clearly that $h(x) \in H(x,\delta)$ holds for all $x \in X$. In fact, the function $h(x)$ constitutes part of the boundary of the graph of H over a wide range of the state space. One can also easily verify the properties listed in Lemma 4.9. For $\delta = 1$ the boundary of the graph looks very rugged which reflects the sensitive dependence on initial conditions and the erratic dynamics of the logistic map.[12] For smaller values of δ this feature disappears because of the smoothing effect of averaging with fast decreasing weights. In the limit as δ approaches zero the average correspondence is as smooth as the function h itself (see part ii) of Lemma 4.9). We can also see clearly that there exist non-trivial fixpoints of H at all depicted δ values. For $\delta = 0.5$ one of these fixpoints is $x = 1/2$. This fixpoint is responsible for the bound $\underline{\delta}(\mathbf{F}) < 1/2$ which we derived from Lemma 4.6. For smaller values of δ this non-trivial fixpoint vanishes and at $\delta = 0.475$ there seems to be only one non-trivial fixpoint left, namely, $x = [1 + \cos(3\pi/8)]/2$. It is the one we have used in the sequel of Equation (4.26) to conclude that the logistic map cannot be optimal for discount rates smaller than 110%. A further reduction of δ makes also this fixpoint vanish and we see therefore that it is not possible to obtain a better bound than $\underline{\delta}(\mathbf{F}) < 0.475$ by applying Theorem 4.2 to the first iterate h. The sharper bound $\underline{\delta}(\mathbf{F}) < 0.25$ was derived from Theorems 4.3 and 4.4 by analysing the higher iterates of h.

4.4 Local Conditions

The Minimum Impatience Theorems and their corollaries presented in Section 4.1 and 4.2 require the evaluation of sufficiently many iterates of h at some state vector $x \in X$. The conditions involved in these results are global in nature since we have to know the policy function h at the points x, $h(x)$, $h^{(2)}(x)$, and so on. In the present section, we develop conditions which can be used to preclude the optimality of the function h in all strictly concave optimal growth problems with a low rate of impatience by utilizing only local properties of h at one particular point $\bar{x}$. To this end, however, we have to assume that h is continuously differentiable in a neighborhood of this point.

We shall derive two different types of local conditions from Theorem 4.2 and Theorem 4.4, respectively. To begin with let us briefly explain the idea on which the local conditions of the first type are based. All we need for an application of the Minimum Impatience Theorem 4.2 is a non-trivial fixpoint of

---

[12]Note, however, that the correspondence depicted in Figure 4.1 is nevertheless continuous.

the average correspondence H. From Lemma 4.9 it follows that every fixpoint of the mapping h is also a fixpoint of the correspondence H, albeit not a non-trivial one. However, if we were able to show that the average correspondence "expands" sufficiently fast in the neighborhood of a trivial fixpoint $\bar{x}$, then it would follow that there exist also non-trivial fixpoints close to $\bar{x}$.

As an illustration consider Figure 4.1 in the previous section. The state $\bar{x} = 0.75$ is a fixpoint of h and, hence, also of H. Moreover, it is evident from the figure that both the upper boundary and the lower boundary of the graph of H are infinitely steep at $\bar{x}$. Therefore, it is clear that there exist non-trivial fixpoints of H close to $\bar{x}$. The situation remains the same in Figure 4.2, although the infinite steepness of the upper boundary is no longer visible because of the limited plotting accuracy. In Figures 4.3 and 4.4, on the other hand, it appears as if there were no non-trivial fixpoints in a sufficiently small neighborhood of $\bar{x}$ because the average correspondence "expands" too slowly. The conclusions drawn from the inspection of Figures 4.1 - 4.4 will be confirmed by Theorem 4.6 below. More specifically, we shall prove that the boundaries of the graph of the average correspondence $H(x, \delta)$ are infinitely steep at $\bar{x}$ for all $\delta > 0.5$.

Now consider the second fixpoint of the logistic map, that is, the point $\tilde{x} = 0$. Scrutinizing Figures 4.1 and 4.2 we see again that the boundaries of the graph of H are steep. This time, however, there are no non-trivial fixpoint close to $\tilde{x}$, because the entire graph of H is located above the diagonal. The average correspondence is expanding but not in a proper way. It will be seen in the following that the different behavior of the average correspondence at $\bar{x}$ and $\tilde{x}$ results from the fact that h is decreasing at $\bar{x}$ whereas it is increasing at $\tilde{x}$.

To grasp the ideas discussed above we now introduce δ-expanding matrices and δ-expanding fixpoints. If $x \in \mathbf{R}^n$ and $\gamma > 0$, then we denote by $B(x, \gamma)$ the open ball in $\mathbf{R}^n$ with center x and radius γ.

Definition 4.3 Let J be a real $n \times n$ matrix, $C \subseteq \mathbf{R}^n$ a given non-empty set, and $\delta \in (0, 1)$ a real number. The matrix J is said to be *δ-expanding* relative to the set C if there exists a vector $u \in C$ such that the following is true: for all real numbers $\gamma > 0$ there exists a number $M(\gamma) \in \mathbf{Z}^+$ such that

$$B(0, \gamma) \subseteq \operatorname{conv}\left\{\delta^m J^m u \,\middle|\, m = 1, 2, \ldots, M(\gamma)\right\}.$$

We shall need the following result.

Lemma 4.10 *Let J be a real $n \times n$ matrix, $C \subseteq \mathbf{R}^n$ a non-empty set, and $\delta \in (0, 1)$ a real number. If the matrix J is δ-expanding relative to C and if λ is an eigenvalue of J, then it holds that $|\lambda| > 1/\delta$. In particular, the matrices J and $I - \delta J$ are non-singular, where I denotes the unit matrix of dimension n.*

PROOF. Assume that λ is an eigenvalue of J with $|\lambda| \leq 1/\delta$ and let Λ be the corresponding eigenspace. If $u \in C$ is the vector mentioned in Definition 4.3, then we denote by u_Λ its orthogonal projection on Λ. The orthogonal projection of $\delta^m J^m u$ onto the subspace Λ is given by $(\delta\lambda)^m u_\Lambda$ and remains uniformly bounded for all $m \in \mathbf{Z}^+$ on account of $|\lambda| \leq 1/\delta$. Therefore, the ball $B(0, \gamma)$ cannot be contained in the convex hull of $\{\delta^m J^m u \mid m \in \mathbf{Z}^+\}$ when γ is chosen sufficiently large. This completes the proof. $\qquad\square$

For the definition of a δ-expanding fixpoint we need the following notation. Let $X \subseteq \mathbf{R}^n$ be the state space and $\bar{x} \in X$ a feasible point. The set

$$C(X, \bar{x}) = \left\{ y \in \mathbf{R}^n \,\middle|\, \exists \epsilon_0 > 0 \text{ such that } \bar{x} + \epsilon y \in X \text{ for all } \epsilon \in (0, \epsilon_0) \right\} \quad (4.35)$$

is called the *cone of interior directions* of X at $\bar{x}$. It is easy to see that this set is a non-empty convex cone. Moreover, $C(X, \bar{x})$ is equal to the unique linear subspace which is parallel to the affine hull of X whenever $\bar{x}$ is in the relative interior of X.

Definition 4.4 Let $X \subseteq \mathbf{R}^n$ be a state space, $h : X \mapsto X$ an arbitrary function, and $\delta \in (0, 1)$ a real number. A vector $\bar{x} \in X$ is called a δ-*expanding* fixpoint of h if the function h is continuously differentiable in a neighborhood[13] of $\bar{x}$ and if the Jacobian matrix $J = h'(\bar{x})$ of h at $\bar{x}$ is δ-expanding relative to the cone of interior directions $C(X, \bar{x})$.

We can now state the first main result of this section.

Theorem 4.6 *Let $X \subseteq \mathbf{R}^n$ be a non-empty, closed, and convex set and let $h : X \mapsto X$ be a given function. If $\bar{x} \in X$ is a δ-expanding fixpoint of h, then there exists a non-trivial fixpoint of the average correspondence H at δ. Consequently, there does not exist any strictly concave optimal growth model (T, A) with state space X such that h is the optimal policy function for this model and such that the globally minimal discount factor of A satisfies $\underline{\delta}(\mathbf{F}) \geq \delta$.*

PROOF. In view of Lemma 4.8 it is sufficient to show that there exists a non-trivial fixpoint of the correspondence H_m at δ for some $m \in \mathbf{Z}^+$. To this end recall that $H_m(x, \delta)$ is the convex hull of the points $g_t(x, \delta)$, $t = 1, 2, \ldots, m$, defined in (4.30). Because of our differentiability assumptions we have for all $u \in C(X, \bar{x})$

$$g_m(\bar{x} + \epsilon u, \delta) = g_m(\bar{x}, \delta) + \epsilon g_m'(\bar{x}, \delta)u + o(\epsilon),$$

---

[13]Of course, this neighborhood is understood with respect to the relative topology on X.

where $g'_m(x, \delta)$ denotes the Jacobian of the function $x \mapsto g_m(x, \delta)$ at the point $\bar{x}$. Since $\bar{x}$ is a fixpoint, we have $g_m(\bar{x}, \delta) = \bar{x}$ and

$$g'_m(\bar{x}, \delta) = \frac{1 - \delta}{\delta(1 - \delta^m)} \sum_{t=1}^{m} \delta^t J^t$$

where $J = h'(\bar{x})$. Therefore, we obtain

$$g_m(\bar{x} + \epsilon u, \delta) = \bar{x} + \epsilon V_m u + o(\epsilon) \tag{4.36}$$

where the matrix V_m is given by

$$V_m = \frac{1 - \delta}{1 - \delta^m} \sum_{t=1}^{m} \delta^{t-1} J^t = \frac{1 - \delta}{1 - \delta^m}(I - \delta J)^{-1} J - \frac{1 - \delta}{1 - \delta^m}(I - \delta J)^{-1} \delta^m J^{m+1}.$$

Using the facts that the Jacobian J is δ-expanding relative to $C(X, \bar{x})$ and that the matrices J and $(I - \delta J)^{-1}$ are non-singular (see Lemma 3.8) it is straightforward to verify that there exists a vector $\bar{u} \in C(X, \bar{x})$ such that for all $\gamma > 0$ one can find a number $M(\gamma) \in \mathbf{Z}^+$ such that

$$B(0, \gamma) \subseteq \text{conv}\left\{ V_m \bar{u} \,\middle|\, m = 1, 2, \ldots, M(\gamma) \right\}.$$

Now let us choose $\gamma > |\bar{u}| + 1$. Then we have $B(\bar{u}, 1) \subseteq B(0, \gamma)$ and therefore

$$\begin{aligned} B(\bar{x} + \epsilon \bar{u}, \epsilon) &= \bar{x} + \epsilon B(\bar{u}, 1) \\ &\subseteq \bar{x} + \epsilon B(0, \gamma) \\ &\subseteq \text{conv}\left\{ \bar{x} + \epsilon V_m \bar{u} \,\middle|\, m = 1, 2, \ldots, M(\gamma) \right\}. \end{aligned}$$

Together with (4.36) this implies that we can choose a number $\epsilon > 0$ sufficiently small such that

$$\bar{x} + \epsilon \bar{u} \in \text{conv}\left\{ g_m(\bar{x} + \epsilon \bar{u}, \delta) \,\middle|\, m = 1, 2, \ldots, M(\gamma) \right\} = H_{M(\gamma)}(\bar{x} + \epsilon \bar{u}, \delta).$$

Because of Lemma 4.8 this shows that $\bar{x} + \epsilon \bar{u}$ is a fixpoint of H at δ for all sufficiently small ϵ, say, $\epsilon \in (0, \bar{\epsilon})$. If all of these fixpoints of the average correspondence H were also fixpoints of the policy function h, then we would have $h(\bar{x} + \epsilon \bar{u}) = \bar{x} + \epsilon \bar{u}$ for all $\epsilon \in (0, \bar{\epsilon})$. This, in turn, would require that $\lambda = 1$ is among the eigenvalues of J. Because this is a contradiction to Lemma 4.10 we conclude that there are (infinitely many) non-trivial fixpoints of H at δ. This completes the proof. $\qquad \square$

Theorem 4.6 raises the question of which fixpoints of h are δ-expanding. Instead of providing a general characterization of such fixpoints we merely present some illustrative examples. First of all, if the Jacobian matrix J has a positive real eigenvalue, then it cannot be δ-expanding for any $\delta \in (0,1)$.[14] Together with Lemma 4.10 this implies that all eigenvalues of J have to be located outside the circle with center 0 and radius $1/\delta$ in the complex plane and not on the positive real axis. For the one-dimensional case, $n = 1$, it follows that $\bar{x}$ is a δ-expanding fixpoint of the function h if and only if h is downward sloping at $\bar{x}$ with $h'(\bar{x}) < -1/\delta$. In the two-dimensional case, a little thought reveals that a δ-expanding fixpoint requires two complex conjugate eigenvalues with non-zero imaginary parts and absolute values greater than $1/\delta$. In the case $n = 3$, one eigenvalue has to be real and smaller than $-1/\delta$ and the other two eigenvalues have to be complex conjugate with the same properties as in the case $n = 2$. One can proceed in this way to discuss also higher dimensional cases. The general conclusion we can draw from this discussion is that the local behavior of the trajectories at the fixpoint $\bar{x}$ must be unstable as well as non-monotonic in every possible direction. This confirms an observation already made before, namely, that the driving force for the Minimum Impatience Theorems 4.1 and 4.2 is the non-monotonicity of optimal growth paths.

Some examples might help to illustrate the above discussion and to demonstrate the usefulness of Theorem 4.6. Both the logistic map $h(x) = 4x(1 - x)$ and the tent map $h(x) = 1 - |2x - 1|$ have a fixpoint $\bar{x}$ with $h'(\bar{x}) = -2$. This implies that $\bar{x}$ is δ-expanding for all $\delta > 1/2$ and it follows that the globally minimal discount factor of any strictly concave optimal growth problem having one of these two maps as its optimal policy function cannot be greater than $1/2$. Of course, we already know this result from Section 4.2 but this time we have used for its derivation only the simple property that the slope of the policy function at the fixpoint $\bar{x}$ is equal to -2. Now consider the generalized tent maps $h_c : [0,1] \mapsto [0,1]$, $c \in (0,1)$, defined in (4.27). It is easy to see that the slope of h_c at the fixpoint $\bar{x} = (2 - c)^{-1}$ is given by $h'_c(\bar{x}) = -(1 - c)^{-1}$ which implies that $\bar{x}$ is δ-expanding for all $\delta > 1 - c$. Therefore, Theorem 4.6 gives us the estimate $\underline{\delta}(\mathbf{F}) \leq 1 - c$ which is basically the same as the one derived from Lemma 4.6, only this time we have obtained it from local information on h_c only.

Next we consider the two-dimensional examples discussed in Section 4.1. As for the Hénon map (4.16) we cannot apply Theorem 4.6 as both fixpoints are saddles with one positive and one negative eigenvalue and we have already indicated that positive eigenvalues preclude δ-expansiveness for any $\delta \in (0,1)$. On the other hand, the mapping defined in (4.17) has a unique fixpoint $(\bar{x}, \bar{y}) = (0.20542, 0.20542)$ which is easily shown to be 0.805-expanding. As in the case

---

[14]This is so because a positive real eigenvalue implies locally monotonic trajectories in the direction of the corresponding eigenvector. The reader should compare this argument with the one put forward on p. 61 concerning monotonic trajectories.

of the tent map, we obtain a bound on the minimal discount factor from the local Minimum Impatience Theorem 4.6 which is not quite as good as the bound obtained from the global Minimum Impatience Theorem 4.1.

We proceed by demonstrating how Theorem 4.6 can be used to exclude the optimality of the function h in strictly concave optimal growth models with low time-preference rates by analysing periodic points of arbitrary period $\tau \in \mathbf{Z}^+$. Because a periodic point with period $\tau \geq 2$ is, in general, not a fixpoint of the average correspondence (see, e.g., Figure 4.4), one would think that it is impossible to apply a similar technique as the one used in Theorem 4.6. However, one can reduce the case of a τ-periodic point to that of a fixpoint by considering an auxiliary optimal growth problem obtained via the time-τ transformation. The τ-periodic point is thereby transformed into a fixpoint of the optimal policy function of the auxiliary problem. This idea is made precise in the following result.

Theorem 4.7 *Let $X \subseteq \mathbf{R}^n$ be a state space, $h : X \mapsto X$ an arbitrary continuously differentiable function, and $\delta \in (0,1)$ a real number. Assume that h has a τ-periodic point $\bar{x} \in X$ and denote by $J_i = h'(h^{(i-1)}(\bar{x}))$, $i = 1, 2, \ldots, \tau$, the Jacobian matrices of h at the iterates of $\bar{x}$. If the matrix*

$$J = J_\tau J_{\tau-1} \cdots J_1$$

is δ^τ-expanding relative to the cone of interior directions $C(X, \bar{x})$, then there does not exist a strictly concave optimal growth model (T, A) with state space X and with a globally minimal discount factor $\underline{\delta}(\mathbf{F})$ greater than or equal to δ such that h is the optimal policy function of (T, A).

PROOF. The result follows easily from Theorems 4.3 and 4.6 by noting that the periodic point $\bar{x}$ is a fixpoint of the τ-th iterate of h and that the Jacobian matrix of $h^{(\tau)}$ at the point $\bar{x}$ is given by J. □

As an example for the application of this result we consider once more the generalized tent map $h_c : [0,1] \mapsto [0,1]$ defined for all parameters $c \in (0,1)$ in (4.27). We have already seen that h_c requires a minimal discount factor smaller than $1 - c$ for being optimal in a strictly concave optimal growth model. This is a very poor result if c is close to zero. By using Theorem 4.7 we can improve the bound $\underline{\delta}(\mathbf{F}) < 1 - c$ for all $c < 1/2$ as follows. The point $\bar{x} = c(1 + c - c^2)^{-1}$ is a periodic point of h_c with minimal period 2. The slope of h_c at $\bar{x}$ and $h_c(\bar{x})$ is given by $1/c$ and $-1/(1 - c)$, respectively. To apply Theorem 4.7, we have to consider the matrix

$$J = J_2 J_1 = \left(-\frac{1}{c(1 - c)} \right)$$

which is δ-expanding for all $\delta > c(1 - c)$. This shows that the mapping h_c cannot be optimal in any strictly concave optimal growth model with a globally minimal discount factor greater than $\sqrt{c(1-c)}$. Since $\sqrt{c(1-c)} < 1 - c$ for all $c \in (0, 1/2)$, it follows that this bound is better than the previous one for small parameter values c.

Although the above examples show that the conditions of Theorem 4.6 can be verified in a number of examples, it has to be admitted that δ-expansiveness is a rather strong assumption. In particular, if the dimension of the state space is high, the requirement that all eigenvalues should have absolute value greater than 1 is certainly restrictive. In the rest of this section we shall demonstrate that another local version of the Minimum Impatience Theorems can be derived from Theorem 4.4 which requires only one eigenvalue to have absolute value greater than 1. Unfortunately, we have to impose another assumption, namely, that this eigenvalue is real and has multiplicity one.

Theorem 4.8 *Let $X \subseteq \mathbf{R}^n$ be a state space with non-empty interior, $h : X \mapsto X$ a given function on X, and $\bar{x} \in X$ a fixpoint of h. Assume that $\bar{x} \in \text{int } X$, that h is continuously differentiable in a neighborhood of $\bar{x}$, and that the Jacobian $J = h'(\bar{x})$ has the positive real eigenvalue λ with algebraic multiplicity one. Then there does not exist a strictly concave optimal growth model (T, A) on the state space X with a globally minimal discount factor $\underline{\delta}(\mathbf{F})$ greater than $1/|\lambda|$ such that h is the optimal policy function of (T, A). If $\lambda < 0$ then the result remains true provided that $-\lambda$ is not an eigenvalue of J.*

PROOF. There is nothing to prove if $|\lambda| \leq 1$. We may therefore assume that either $\lambda > 1$ or $\lambda < -1$.

Case 1: $\lambda > 1$. We claim that for any real number $\eta > 0$ there exist a number $\delta \in (1/\lambda - \eta, 1/\lambda + \eta)$ and a vector $y \in \mathbf{R}^n$ such that $y \neq 0$, $\bar{x} + y \in X$, and

$$h(\bar{x} + \delta y) = \bar{x} + y. \tag{4.37}$$

If this claim can be proved, then the result follows from Theorem 4.4 by setting $\tau = 1$, $\bar{x}_0 = \bar{x}$, and $x_0 = \bar{x} + \delta y$. To verify the claim we assume that all coordinates are with respect to a real basis $\{e_1, e_2, \ldots, e_n\}$ where e_1 is an eigenvector corresponding to λ.[15] In particular, the vector y in Equation (4.37) has the representation $y = (y_1, y_2, \ldots, y_n)$. Since $\bar{x}$ is a fixpoint of h we can rewrite Equation (4.37) as

$$\lambda \delta y_1 + \Phi_1(\delta y_1, \delta y_2, \ldots, \delta y_n) = y_1, \tag{4.38}$$

$$\tilde{J} \begin{pmatrix} \delta y_2 \\ \vdots \\ \delta y_n \end{pmatrix} + \begin{pmatrix} \Phi_2(\delta y_1, \delta y_2, \ldots, \delta y_n) \\ \vdots \\ \Phi_n(\delta y_1, \delta y_2, \ldots, \delta y_n) \end{pmatrix} = \begin{pmatrix} y_2 \\ \vdots \\ y_n \end{pmatrix}, \tag{4.39}$$

---

[15]Such a basis always exists since λ is a real eigenvalue.

where the functions $\Phi_j(z_1, z_2, \ldots, z_n)$, $j = 1, 2, \ldots, n$, satisfy the conditions

$$\Phi_j(0, 0, \ldots, 0) = 0$$

and

$$\frac{\partial}{\partial z_i} \Phi_j(0, 0, \ldots, 0) = 0$$

for all $i = 1, 2, \ldots, n$ and where $\tilde{J}$ is a $(n-1) \times (n-1)$ real matrix such that $\tilde{J} - \lambda I$ is non-singular. The latter property implies that we can solve Equation (4.39) for δy_i, $i = 2, 3, \ldots, n$ locally around $y_i = 0$ and $\delta = 1/\lambda$ to obtain

$$\delta y_i = g_i(y_1, \delta) \qquad i = 2, 3, \ldots, n. \tag{4.40}$$

From (4.39) and $\Phi_j(0, 0, \ldots, 0) = 0$, $j = 2, 3, \ldots, n$, it follows that these solutions satisfy $g_i(0, \delta) = 0$ for all $i = 2, 3, \ldots, n$ and for all $\delta \in B(1/\lambda)$, where $B(1/\lambda)$ is a sufficiently small neighborhood of $1/\lambda$. Substituting (4.40) into (4.38) we obtain the equation

$$f(y_1, \delta) = (\lambda \delta - 1)y_1 + \Phi_1(\delta y_1, g_2(y_1, \delta), \ldots, g_n(y_1, \delta)) = 0.$$

Finally, by noting that $f(0, \delta) = 0$ for all $\delta \in B(1/\lambda)$ and that $\partial f(0, \delta)/\partial y_1 = \lambda \delta - 1$ it is easily seen that for every $\epsilon > 0$ and for every $\eta > 0$ one can find δ in the η-neighborhood of $1/\lambda$ and a number $y_1(\delta) \neq 0$ such that $f(y_1(\delta), \delta) = 0$ and $|y_1(\delta)| < \epsilon$. This shows that the vector $y = (y_1, y_2, \ldots, y_n)$ defined by $y_1 = y_1(\delta)$ and $y_i = g_i(y_1(\delta), \delta)$, $i = 2, 3, \ldots, n$, solves Equations (4.38) and (4.39). Since $|y_1(\delta)|$ can be chosen to be smaller than an arbitrary but fixed positive number ϵ and since the functions g_i are continuous it follows that the vector y can be made arbitrarily small (in norm). Together with the assumption that $\bar{x} \in \text{int } X$ this implies that $\bar{x} + y \in X$ and the claim is proven.

Case 2: $\lambda < -1$. This case can be reduced to Case 1 by considering the second iterate of h. Indeed, if h is the optimal policy function of (T, A), then there exists another strictly concave optimal growth model (T', A') with $h^{(2)}$ as its optimal policy function (see Lemma 3.10). Moreover, $\bar{x}$ is a fixpoint of $h^{(2)}$ and the Jacobian of $h^{(2)}$ at $\bar{x}$ has the real and positive eigenvalue λ^2 with multiplicity one (since $-\lambda$ is not an eigenvalue of J). Therefore, it follows from Case 1 that $\underline{\delta}(\mathbf{F}') \leq 1/\lambda^2$ which in turn implies $\underline{\delta}(\mathbf{F}) \leq 1/|\lambda|$ by Lemma 3.10 ii). This completes the proof. $\qquad \square$

Two remarks are in order concerning the assumptions of Theorem 4.8. First of all, the requirement that the algebraic multiplicity of λ has to be one is crucial for the proof which is based on Theorem 4.4. To see this consider the mapping $h : \mathbf{R}^2 \mapsto \mathbf{R}^2$ defined by

$$(x, y) \mapsto (2x + y(x^2 + y^2), \; 2y - x(x^2 + y^2)).$$

This mapping has the fixpoint $(0,0)$ with the double eigenvalue $\lambda = 2$. On the other hand, it is easy to verify that the equation $(x, y) = \delta h(x, y)$ does not have a non-trivial solution for any $\delta \in \mathbf{R}$ so that Theorem 4.4 cannot be applied.

Analogously, one can see that complex conjugate eigenvalues cannot be handled by our proof. As a matter of fact, we have seen in Section 4.1 that the mapping h defined by (4.17) has a unique fixpoint $\bar{x}$ but that the equation $x - \bar{x} = \delta(h(x) - \bar{x})$ does not have a solution $x \neq \bar{x}$ for any $\delta \in \mathbf{R}$.

Let us conclude this section by some examples. First consider the Hénon map (4.16). Since both of its fixpoints have only real eigenvalues, we could not apply Theorem 4.6. The assumptions of Theorem 4.8, on the other hand, are satisfied and we obtain the same bound $\underline{\delta}(\mathbf{F}) < 0.307$ which we have already proved in Section 4.1 (see p. 72). More specifically, one of the eigenvalues of the Jacobian evaluated at the fixpoint $(-1.13135, -1.13135)$ is given by $\lambda = 3.25982$ so that Theorem 4.8 gives $\underline{\delta}(\mathbf{F}) \leq 1/\lambda \approx 0.307$.

As a second example let us consider one-dimensional maps in general. We have seen earlier that $1/|h'(\bar{x})|$ is a bound on the minimal discount factor provided that h is negatively sloped at the fixpoint $\bar{x}$. Under the assumption that $\bar{x}$ is an interior fixpoint we can use Theorem 4.8 to obtain the same result without the requirement of a negative slope.

4.5   Open Questions

As every new result the Minimum Impatience Theorems raise questions about their implications and possible extensions. We are going to discuss some of these issues in the present section.[16]

It has already been emphasized that the proofs of our central results, Theorem 4.1 and Theorem 4.4, depend heavily on the dynamic programming conditions of Theorem 3.1. In fact, the geometric properties of the functions G and W stated in Theorem 3.1 have been identified as the foundation on which all Minimum Impatience Theorems rest. This suggests that we can expect similar results to hold as long as an analogue to Theorem 3.1 is available. This will be the case for most other deterministic optimal growth models which are formulated in terms of recursive utility functionals even if the assumptions on the primitives (T, A) differ from those stated in Definitions 2.1 and 2.3. There are such models and we have already commented on more general sets of assumptions at the end of Section 2.2. Moreover, we have noted that one advantage of our formulation is that certain uniform summability conditions are automatically satisfied. These conditions were used in the proof of Theorem 4.1 when

---

[16]In a logical sense, this can be regarded as the final section of the entire book since the material presented here applies equally well to the continuous time framework of Part 2.

we changed the order of summation in various infinite series. The summability conditions, however, are not indispensable for the proof since we could always avoid the infinite series by allowing only finitely many of the weights μ_t in conditions (4.1) - (4.3) to be non-zero. This would yield the same Minimum Impatience Theorems except that the average correspondence H would have to be replaced by the approximating sequence of correspondences H_1, H_2, H_3, ... introduced in Section 4.3.

The next question regards the possible improvements of our results under the present assumptions. More specifically, we would like to know how tight the bounds are which we derived from the Minimum Impatience Theorems. To get a rough idea on the answer to this question let us summarize in the following table the results for those of our examples for which strictly concave optimal growth models have explicitly been constructed.

mapping	best model	best bound	source
logistic map	$\delta < 0.041\dot{6}$	$\delta \le 0.25$	Lemma 3.11 and p. 77
tent map	$\delta < 0.25$	$\delta < 0.445$	Lemma 3.12 and p. 75
Hénon map	$\delta = 0.2$	$\delta < 0.307$	[14, p. 201] and p. 72

This table indicates that our sharpest bound is roughly 1.5 – 6 times larger than the highest value which is reported in the literature and which is known to the author. Of course, one should be very cautious in generalizing these results since the examples might not be representative. All mappings listed in the above table are chaotic and it is not clear whether the Minimum Impatience Theorems work as well for less complicated policy functions as they do for chaotic ones. In this regard we would like to point out once more that the Minimum Impatience Theorems derived from Theorem 4.1 require non-monotonicity in every possible direction.[17] For example, these results do not provide a useful bound $\delta \in (0,1)$ for any two-dimensional policy function h of the form

$$h(x,y) = (h_1(x,y), h_2(x,y))$$

where h_1 and h_2 are arbitrary continuous functions with $h_2(x,y) > y$ for all (x,y) in the state space X. This simply follows from the fact that the trajectories of h are strictly monotonic in the y direction. However complicated the dynamical behavior of the x component is, we cannot conclude from Theorem 4.1 or one of

---

[17]The results derived from Theorem 4.4, on the other hand, do not require non-monotonic growth paths.

its corollaries that this policy function is not optimal because of the simple (i.e., monotonic) dynamics of the y component. To summarize, we believe that there is still ample room for improvements of the Minimum Impatience Theorems but the methods required for these improvements might be very different from those employed in the present book.

Now let us ponder upon the question whether the Minimum Impatience Theorems would remain valid, at least in principle, if we allowed for uncertainty in the optimal growth models. So far we have not been able to prove a useful stochastic version of any of our results, but there is some reason to believe that the answer should be in the affirmative. To explain this point in more detail, let us consider the following standard scenario of stochastic economic growth: a decision maker with preferences described by an aggregator function A faces the constraints of a technology T with the additional feature that she/he cannot determine the next period's capital stock $x' = h(x)$ exactly but only a probability distribution $Q(x, \cdot)$ from which the actual capital stock x' is drawn. Dynamic programming works in this model just as well as it did in the deterministic model and, as we have identified dynamic programming as the basic tool for proving the Minimum Impatience Theorems, a proof of this result along similar lines as in the deterministic case might be possible. Moreover, it appears to us that the intuitive idea of the Minimum Impatience Theorems is rather robust. It shows that a decision maker reveals her/his degree of impatience by the way she/he acts and there does not seem to be any reason why this should fail to be true if the model includes uncertainty. Finally, it does not seem to be very likely that a growth path, which fluctuates too wildly to become optimal in any strictly concave deterministic growth problem with a discount factor $\underline{\delta}(D) \geq 0.2$, should turn out to be optimal in some stochastic model with a discount factor of, say, $\underline{\delta}(D) = 0.5$ when the random noise in that model is very small as compared to the variation of the given path. This last statement makes at least three important points. First, it is clear that we cannot expect uncertainty to have no impact at all on the bound on the discount factor. A stochastic Minimum Impatience Theorem would probably yield a higher bound on the discount factor than its deterministic counterpart if both were applied to the same growth path. Second, a stochastic version of the theorem would not provide any useful information if the uncertainty had such a strong effect that it would drown the decision makers efforts to steer the economy. Third, the stochastic Minimum Impatience Theorems would probably not provide a sharp bound like $\delta < \delta^h$ but only some probabilistic bound. For example, a result in terms of a confidence interval would sound intuitively plausible. In this regard, it should be noted that in the stochastic model described above the optimal growth paths $_0x$ are realizations of a stochastic process and that the probability law of this process depends on the parameters $\underline{\delta}(D)$, $D \subseteq \mathbf{F}$. The existence of confidence intervals for these parameters would not be totally surprising although their derivation might be very difficult.

We have motivated the Minimum Impatience Theorems by the recent interest in finding necessary conditions for the occurrence of *chaotic* dynamics in normative economic models. Strictly speaking, however, this issue has not been addressed at all in this book. In fact, a good bound on the discount factor obtained from the Minimum Impatience Theorems does not necessarily imply that the policy function h under consideration is chaotic[18] and in order to decide whether chaotic behavior of h provides a sufficient condition for the existence of a non-trivial bound $\delta < 1$ one would have to give the phrase "chaotic behavior" a precise mathematical meaning. This, however, can be done in quite a few different ways. Among the possible chaos definitions are the sensitive dependence on initial conditions, the existence of a positive Ljapunov exponent, the existence of an invariant ergodic measure (with some additional property like absolute continuity with respect to the Lebesgue measure), and the strict positivity of the Kolmogorov entropy (see, e.g., [31] and [29]). To obtain something which is close to a necessary optimality condition for chaos one would need a new type of Minimum Impatience Theorems taking into account one or more of the aforementioned indicators of chaotic dynamical behavior. Let us illustrate this point in more detail by considering the chaos-definition in terms of the existence of a positive Ljapunov exponent. The local conditions presented in Section 4.4 relate the eigenvalues of a fixpoint (or a periodic point) of the given policy function h to the minimum rate of impatience required for the optimality of h in any strictly concave optimal growth model. Ljapunov exponents can be regarded as generalizations of eigenvalues to non-stationary trajectories and, hence, it is not completely unrealistic to conjecture a relation between the Ljapunov exponents of h and the minimum rate of impatience. This could be as simple a relation as $\underline{\delta}(\mathbf{F}) < e^{-\lambda}$, where λ denotes the largest Ljapunov exponent, or it could be a more complicated one.[19] At any rate, it would provide quite a remarkable generalization of Theorems 4.6 and 4.7 and

---

[18]In Part 1 of the book we have applied the Minimum Impatience Theorems only to mappings with extremely complicated dynamical behavior. However, it is easy to come up with examples with less complicated dynamics for which our results indicate a very small minimal discount factor. Just think of a strictly decreasing function $h : [0,1] \mapsto [0,1]$ and make it sufficiently steep at its (unique) fixpoint: Theorem 4.6 yields a small bound on the minimal discount factor. The continuous time example discussed in Section 7.1 will also show that good bounds can be obtained for dynamical systems with relatively simple dynamics.

[19]Note that the conjecture $\underline{\delta}(\mathbf{F}) < e^{-\lambda}$ is also motivated by Theorem 4.4, since the Ljapunov exponents measure the sensitivity with respect to initial conditions averaged over all possible initial conditions. A somewhat bolder conjecture would be a relation between the minimal discount factor $\underline{\delta}(\mathbf{F})$ and the sum of all positive eigenvalues, i.e., the Kolmogorov entropy (see [50] for a first result in that direction).

– this is at least our conjecture – it would be about as close to a necessary optimality condition for chaos as one could possibly get. As a matter of fact, we do strongly believe that chaos can be optimal at arbitrary small discount rates, although the optimal paths will look less chaotic to the naked eye when the discount rates are getting close to zero.[20]

---

[20]Partial substantiation for this conjecture in the continuous time framework comes from a result by Montrucchio [49] who demonstrated that "slow chaos", that is, chaos with only small positive Ljapunov exponents can be optimal for arbitrary small discount rates.

Part II

Continuous Time Models

Chapter 5

Model Formulation

The second part of this book deals with exactly the same issue as the first one except that we now consider recursive dynamic models in a continuous time framework. Because of this reason we shall restrict the presentation in this part to mathematical aspects for the most time and we shall add economic interpretations only when they are different or new as compared to the discrete time models. Moreover, we shall frequently refer back to Part 1 when certain arguments or complete proofs are similar to the corresponding portions in the discrete time case. In this regard, we caution the reader that most of the notations used in Part 2 are identical to the corresponding terms in Part 1 although there is a different meaning attached to them now. For example, when we talk about the β-topologies in the continuous time framework, one has to be aware of the fact that these topologies are similar but not identical to the β-topologies introduced in Section 2.1. We hope that using the same or similar notations in both parts helps to simplify the presentation rather than to cause confusion. In any case, we assume that the reader has already completed Part 1 of the book when she/he starts to work through Part 2.

In the present chapter we define the technology and the preferences for continuous time recursive economic models. Whereas the basic ideas and concepts are the same as in the discrete time case, the technical details and proofs are rather different at some places. For example, the existence of β-myopic recursive utility functionals is not proved by an application of Banach's fixpoint theorem but follows from existence theorems for solutions to ordinary differential equations. This approach was developed by Epstein [30] under somewhat different assumptions than ours.

In Section 5.1 we begin to formulate the model by defining the technology of the economy and we discuss various properties of the induced set of feasible growth paths. In Section 5.2 we derive recursive utility functionals from suitably defined aggregator functions. Although our approach is motivated by [30], there are two important differences to Epstein's work. First, as in the discrete time framework we are aiming at a reduced form optimal growth model in which consumption does not occur explicitly. Second, we try to develop a continuous time analogue of the model set up in Chapter 2 which means that we require the utility functionals to be continuous with respect to the β-topology rather than with respect to the Mackey topology as it is the case in [30]. Finally, in Section 5.3 we integrate the technology and the preferences into the complete model.

Since we consider optimal growth models in a continuous time setting, the time variable t will from now on denote a non-negative real number, i.e., $t \in [0, \infty)$. If $x : [0, \infty) \mapsto S$ is a function with values in some set S we shall denote its value at time t either by $x(t)$ or by x_t. Furthermore, for any number $t \in [0, \infty)$ we define the function $_tx : [0, \infty) \mapsto S$ by $_tx(s) = x(t + s)$ for all $s \in [0, \infty)$ (equivalently, we could write this as $_tx_s = x_{t+s}$). In other words, the function $_tx$ is the concatenation of the right shift (by t) on the non-negative real line and the original function x. As a special case of this notational convention we shall frequently write $_0x$ instead of x in order to distinguish the function $_0x$ from a vector $x \in S$.

5.1   Technology

A feasible growth path in the continuous time optimal growth problem to be developed in this chapter is a continuous function from the time domain $[0, \infty)$ to the set of feasible capital stocks. Let us denote the set of feasible capital stocks by X and let us as in the discrete time setting assume that X is a non-empty, closed, and convex subset of $\mathbf{R}^n$ where $n \in \mathbf{Z}^+$ denotes the number of different capital goods. If $x : [0, \infty) \mapsto X$ is a feasible growth path, then $x_t \in X$ is the vector of capital stocks available in the economy at time t.

In contrast to a discrete time model it is not possible to specify the technological constraints by the set of feasible capital stocks for the next period. Instead, one usually defines the technology by the feasible rates of change of the current capital stock, i.e., by the feasible net investment rates. This approach requires that the net investment rates are well defined for all $t \in [0, \infty)$ or, equivalently, that the feasible growth paths are differentiable functions. On the other hand, it is well known that the class of differentiable functions is quite often too small to contain optimal solutions so that it has become common to consider a larger class of functions, namely, the class of all absolutely continuous functions.[1] Since the derivative of an absolutely continuous function is uniquely

---

[1] A function $x : [0, \infty) \mapsto \mathbf{R}^k$ is said to be absolutely continuous if there exists a measurable and locally integrable function $\dot{x} : [0, \infty) \mapsto \mathbf{R}^k$, called the derivative of $_0x$, such that the equation $x_t = x_s + \int_s^t \dot{x}_\tau \, d\tau$ holds for all $s, t \in [0, \infty)$.

determined only up to modifications on sets of Lebesgue measure zero, it makes sense to describe properties of derivatives only outside an (unspecified) set of measure zero. More specifically, we say that a property P holds almost everywhere (a.e.) on $[0,\infty)$ if there exists a set $E \subseteq [0,\infty)$ with Lebesgue measure zero such that P is true for all $t \in [0,\infty)\backslash E$.

If $S \subseteq \mathbf{R}^k$ is a given set, then we denote by $\mathcal{AC}(S)$ the set of all absolutely continuous functions $x : [0,\infty) \mapsto S$. For a given function $_0x \in \mathcal{AC}(S)$ we denote by $_0\dot{x}$ its derivative, i.e., any function satisfying $x_t = x_s + \int_s^t \dot{x}_\tau \, d\tau$ for all $s,t \in [0,\infty)$.

We assume that the technology is stationary which implies that the technological constraints of the economy can be described by a production technology set $T \subseteq X \times \mathbf{R}^n$. A function $x \in \mathcal{AC}(X)$ is a feasible growth path if the pair $(x_t, \dot{x}_t)$ is an element of T for almost every $t \in [0,\infty)$. As in Part 1 we assume that the set T is closed and convex and that it satisfies a growth condition of the form $|x'| \leq \alpha + \beta|x|$ for all $(x, x') \in T$. Finally, it will be assumed that the x-section of T defined by $T_x = \{x' \in \mathbf{R}^n \,|\, (x, x') \in T\}$ is non-empty for every feasible capital stock $x \in X$.

The above discussion is summarized in the following definitions.

Definition 5.1 Let $n \in \mathbf{Z}^+$ be given. A set $X \subseteq \mathbf{R}^n$ is called a *state space* if it is non-empty, closed, and convex. A set $T \subseteq X \times \mathbf{R}^n$ is said to represent a *technology* on the state space X, if the following conditions are satisfied.

   i) The set T is closed and convex.

   ii) The set $T_x = \{x' \in \mathbf{R}^n \,|\, (x, x') \in T\}$ is non-empty for every $x \in X$.

   iii) There exist real numbers $\alpha \geq 0$ and $\beta \geq 0$ such that the inequality $|x'| \leq \alpha + \beta|x|$ holds for all $(x, x') \in T$.

Definition 5.2 A *feasible growth path* for a given technology T on the state space $X \subseteq \mathbf{R}^n$ is an absolutely continuous function $_0x \in \mathcal{AC}(X)$ such that $(x_t, \dot{x}_t) \in T$ holds for almost every $t \in [0,\infty)$. The set of feasible growth paths will be denoted by $\mathbf{F}$. Moreover, for every feasible state vector $x \in X$ we denote by $\mathbf{F}(x)$ the subset of $\mathbf{F}$ consisting of those feasible growth paths $_0x \in \mathbf{F}$ which satisfy the initial condition $x_0 = x$.

A technology as defined in Definition 5.1 is called a *β-bounded* technology. The following result is identical to Lemma 2.1.

Lemma 5.1 *If T is a technology on the state space $X \subseteq \mathbf{R}^n$, then it holds that the x-section T_x is a non-empty, compact, and convex set for every $x \in X$ and that the set-valued mapping $x \mapsto T_x$ is continuous on X.*

In order to be able to define β-myopic utility functionals we have to endow the set $\mathbf{F}$ with a topology. To this end let us define for every positive real number β the β-norm of a function $_0x \in \mathcal{AC}(\mathbf{R}^n)$ by[2]

$$\|_0x\|_\beta = \sup\left\{|x_t|e^{-\beta t} \,\Big|\, t \in [0,\infty)\right\} + \text{ess sup}\left\{|\dot{x}_t|e^{-\beta t} \,\Big|\, t \in [0,\infty)\right\}.$$

The function space $\mathbf{F}_\beta$ is defined as

$$\mathbf{F}_\beta = \left\{_0x \in \mathcal{AC}(\mathbf{R}^n) \,\Big|\, \|_0x\|_\beta < \infty\right\}.$$

We call the norm topology on $\mathbf{F}_\beta$ the β-topology. It is obvious that convergence of a sequence of functions $_0x^{(k)} \in \mathbf{F}_\beta$ in the β-topology to an element $_0x \in \mathbf{F}_\beta$ implies uniform convergence of $_0x^{(k)}$ on every compact interval and uniform convergence of $_0\dot{x}^{(k)}$ almost everywhere on every compact interval. In other words, the β-topologies are stronger than the topology of uniform convergence on compact intervals (the latter topology is usually referred to as the compact open topology). Moreover, we have the following result.

Lemma 5.2 *Let $\mathbf{F}$ be the set of feasible growth paths in a β-bounded technology T on the state space $X \subseteq \mathbf{R}^n$ and let $x \in X$ be an arbitrary feasible state vector. Then it holds that $\mathbf{F}$ and $\mathbf{F}(x)$ are convex sets. If $\beta > 0$, then $\mathbf{F}$ and $\mathbf{F}(x)$ are subsets of $\mathbf{F}_\beta$ which are closed in the β-topology.*

PROOF. Convexity of $\mathbf{F}$ and $\mathbf{F}(x)$ follows immediately from convexity of X and T. Now assume that $\beta > 0$ and that $_0x \in \mathbf{F}$ is a given feasible growth path. The real-valued function $t \mapsto |x_t|$ is absolutely continuous and satisfies

$$\frac{d}{dt}|x_t| = \begin{cases} \dfrac{x_t \cdot \dot{x}_t}{|x_t|} & \text{if } x_t \neq 0 \\[2mm] 0 & \text{if } x_t = 0 \end{cases}$$

almost everywhere on $[0,\infty)$. This implies

$$\left|\frac{d}{dt}|x_t|\right| \leq |\dot{x}_t| \leq \alpha + \beta|x_t| \qquad \text{a.e..}$$

---

[2]A non-negative and measurable function $y : [0,\infty) \mapsto \mathbf{R}^n$ is said to be essentially bounded if the Lebesgue measure of the set $\{t \,|\, y_t > M\}$ is zero for some real number M. In that case, the essential supremum of $_0y$ is defined by

$$\text{ess sup}\left\{y_t \,\Big|\, t \in [0,\infty)\right\} = \inf\left\{M \,\Big|\, \ell\{t \,|\, y_t > M\} = 0\right\}$$

where ℓ denotes the Lebesgue measure. If $_0y$ is not essentially bounded, then we set $\text{ess sup}\{y_t \,|\, t \in [0,\infty)\} = \infty$.

Applying Gronwall's lemma we obtain from this inequality the estimate

$$|x_t| \leq \left(\frac{\alpha}{\beta} + |x_0| \right) e^{\beta t} - \frac{\alpha}{\beta} . \tag{5.1}$$

Together with part iii) of Definition 5.1 this implies

$$|\dot{x}_t| \leq (\alpha + \beta |x_0|) e^{\beta t} \qquad \text{a.e..}$$

From this inequality and from (5.1) we obtain

$$\|_0 x\|_\beta \leq (1 + \beta) \left(\frac{\alpha}{\beta} + |x_0| \right) < \infty$$

which proves that $_0 x \in \mathbf{F}_\beta$.

Finally, let $_0 x^{(k)} \in \mathbf{F}$ be a feasible growth path for every $k \in \mathbf{Z}^+$ and assume that $_0 x^{(k)}$ converges in the β-topology to an element $_0 \bar{x} \in \mathbf{F}_\beta$ as k approaches infinity. As already pointed out this implies that $\lim_{k \to \infty} x_t^{(k)} = \bar{x}_t$ holds uniformly on compact intervals and $\lim_{k \to \infty} \dot{x}_t^{(k)} = \dot{\bar{x}}_t$ holds uniformly almost everywhere on compact intervals. Since X and T are closed sets, we can conclude that $\bar{x}_t \in X$ for all $t \in [0, \infty)$ and that $(\bar{x}_t, \dot{\bar{x}}_t) \in T$ for almost every $t \in [0, \infty)$. This proves that $_0 \bar{x} \in \mathbf{F}$ and it follows that $\mathbf{F}$ is closed in the β-topology. The closedness of the sets $\mathbf{F}(x)$ for all $x \in X$ can be proved analogously. $\qquad \square$

It should be noted that in contrast to discrete time models the sets $\mathbf{F}$ and $\mathbf{F}(x)$ may well be empty. For example take $X = [0, 1]$ and

$$T = \left\{ (x, x') \,\middle|\, x \in X, \, x \leq x' \leq 1 \right\}.$$

Obviously, the set $\mathbf{F}(x)$ is empty for all $x \in (0, 1]$ and it contains a single element if $x = 0$, namely, the constant path $x_t \equiv 0$. The question of what assumptions have to be imposed on the technology T to guarantee the existence of at least one feasible growth path for every initial state has been answered by the so-called Viability Theorem (see, e.g., [2, 3]). In fact, a necessary and sufficient condition for the existence of feasible paths (also called *viable* trajectories) is that the intersection of T_x with the tangent cone of X at x is non-empty for all $x \in X$.[3] Since existence of feasible solutions is not of primary interest to us we refrain from imposing this additional assumption and simply assume the existence of feasible solutions whenever this is necessary.

---

[3]The tangent cone of X at x is the closure of the cone of interior directions, $C(X, x)$, defined in (4.35).

Finally, note that in the continuous time framework there is no compactness result like Lemma 2.3. To see this, consider the following simple example of a technology which is β-bounded for all $\beta \in [0, \infty)$: $X = \mathbf{R}$ and

$$T = \left\{ (x, x') \,\middle|\, x \in X, x' \in [-1, 1] \right\}.$$

For every $k \in \mathbf{Z}^+$, define the function $_0x^{(k)}$ by $x_0^{(k)} = 0$ and $\dot{x}_t^{(k)} = (-1)^i$ whenever $t \in [(i-1)/k, i/k)$, $i \in \mathbf{Z}^+$. Although it holds that $_0x^{(k)} \in \mathbf{F}$ for all $k \in \mathbf{Z}^+$ there does not exist a subsequence $_0x^{(k_1)}$, $_0x^{(k_2)}$, $_0x^{(k_3)}$, ... which is convergent in the β-topology for some $\beta \geq 0$. Hence, $\mathbf{F}$ cannot be compact in any β-topology.

5.2   Preferences

The next step to be taken is to specify a utility functional on the set $\mathbf{F}$ of feasible growth paths. The properties we would like to be satisfied by such a functional are the same as in the discrete time case: namely, that the functional is continuous with respect to the β-topology (we call this property β-*myopia*) and that the influence of current utility and total future utility is separated in a specific way (we call this property *recursiveness*). Let us start with recursiveness.

In the discrete time case the separation of current and future utility is represented by the recursive equation (2.3) and it has the interpretation that total utility in the current period is determined completely by the total utility in the next period and by the pair (x_0, x_1). Note that a functional dependence on the pair (x_0, x_1) is equivalent to a functional dependence on the current capital stock x_0 and the current net investment $x_1 - x_0$. In continuous time models there is no such thing as a next period and we have to work with rates of change rather than with values in consecutive periods. Therefore, the property of recursiveness in a continuous time model has to be formulated as follows: the rate of change of current utility is a function of the current utility, the current capital stock, and the current net investment rate. Formally, we require that the utility functional $U : \mathbf{F} \mapsto \mathbf{R}$ satisfies the following differential equation[4]

$$\frac{d}{dt} U(_tx) = -A((x_t, \dot{x}_t), U(_tx)) \tag{5.2}$$

for almost all $t \in [0, \infty)$ and for all $_0x \in \mathbf{F}$. Now let us assume that a fixed path $_0x \in \mathbf{F}$ is given and define $u(t) = U(_tx)$. With this specification we see that $U(_0x)$ is the initial value of a solution to the differential equation

$$\dot{u}(t) = -A((x_t, \dot{x}_t), u(t)) \qquad \text{a.e..} \tag{5.3}$$

---

[4]The minus sign in Equation (5.2) corresponds in discrete time models to the fact that (2.3) is a backward recursive equation.

But which of the (in general) infinitely many solutions of this equation should we select as the one that defines $U(_0x)$? We have to find a property which characterizes one of them in a unique way. As already pointed out by Epstein [30] the property which does the job is boundedness. However, because we allow for unbounded growth paths which increase exponentially at a rate smaller than or equal to β, we cannot simply require that u is a bounded function as it was the case in [30]. Instead, the appropriate assumption for solutions to (5.3) is the growth condition

$$\sup\left\{|u(t)|e^{-\beta t}\,\middle|\,t \in [0,\infty)\right\} < \infty. \tag{5.4}$$

It will be shown below that there is indeed a unique solution of Equation (5.3) subject to the condition (5.4) provided that the aggregator function A satisfies some reasonable conditions which are stated in the following definition.

Definition 5.3 Let T be a given β-bounded technology on the state space $X \subseteq \mathbf{R}^n$ and assume that $\beta > 0$. A continuous function $A : T \times \mathbf{R} \mapsto \mathbf{R}$ is called an *aggregator function* for T if the following conditions are satisfied:

i) There exist a constant $M' > 0$ and a vector $\bar{v} \in \mathbf{R}^n$ such that

$$|A((x,x'),\bar{v}\cdot x)| \leq M'(1+|x|)$$

holds for all $(x,x') \in T$.

ii) The function $z \mapsto A((x,x'),z)$ from $\mathbf{R}$ to $\mathbf{R}$ is strictly decreasing and it is locally Lipschitz continuous for all $(x,x') \in T$. Moreover, there exists a real number $\rho > 0$ such that the condition

$$|A((x,x'),z) - A((x,x'),z')| \geq \underline{\rho}|z - z'| \tag{5.5}$$

holds for all $(x,x') \in T$ and for all $z, z' \in \mathbf{R}$.

iii) The parameters β and ρ satisfy the relation $\beta < \underline{\rho}$.

Comparing this definition with its discrete time counterpart, Definition 2.3, one can discover a lot of similarities but also a few differences. In particular, it should be noted that the analogue to the Lipschitz condition (2.5) is the inequality stated in (5.5) and not the assumption of local Lipschitz continuity of the mapping $z \mapsto A((x,x'),z)$. The latter assumption is only imposed for technical reasons and has no counterpart in the discrete time setting.

We can now prove two preliminary results which will be needed to establish the existence of a unique solution to (5.3) and (5.4). The notation used in these results is as follows. The constants α, β, M', and $\underline{\rho}$ as well as the vector $\bar{v}$

are taken from Definition 5.1 and Definition 5.3. Moreover, we define another constant $M > 0$ by

$$M = M' + |\bar{v}| \max\{\alpha, \beta\}$$

and we specify the function $K : X \mapsto \mathbf{R}$ by

$$K(x) = \frac{M}{\underline{\rho} - \beta} \left(1 + \frac{\alpha}{\beta} + |x| \right). \tag{5.6}$$

Lemma 5.3 *Let T be a β-bounded technology on the state space $X \subseteq \mathbf{R}^n$ and assume that $\beta > 0$. Furthermore, let $A : T \times \mathbf{R} \mapsto \mathbf{R}$ be an aggregator function for T and let s and s' be two non-negative real numbers such that $0 \leq s' \leq s$. If $_0x \in \mathbf{F}$ is a feasible growth path and $u : [s', s] \mapsto \mathbf{R}$ is an absolutely continuous function such that both the differential equation (5.3) and the boundary condition $u(s) = \bar{v} \cdot x_s$ are satisfied, then it holds for all $t \in [s', s]$ that*

$$|u(t) - \bar{v} \cdot x_t| \leq K(x_0) \left[e^{\beta t} - e^{\underline{\rho} t - (\underline{\rho} - \beta)s} \right]. \tag{5.7}$$

PROOF. Let us define the function $g : [0, s] \mapsto \mathbf{R}$ as the right hand side of Inequality (5.7), that is,

$$g(t) = K(x_0) \left[e^{\beta t} - e^{\underline{\rho} t - (\underline{\rho} - \beta)s} \right].$$

Note that $g(t) \geq 0$ holds for all $t \in [0, s]$. First, we want to show that u satisfies

$$u(t) - \bar{v} \cdot x_t > g(t) \Rightarrow \dot{u}(t) - \bar{v} \cdot \dot{x}_t > \dot{g}(t) \tag{5.8}$$

and

$$u(t) - \bar{v} \cdot x_t < -g(t) \Rightarrow \dot{u}(t) - \bar{v} \cdot \dot{x}_t < -\dot{g}(t) \tag{5.9}$$

for almost all $t \in [s', s]$. To verify (5.8) we assume $u(t) > g(t) + \bar{v} \cdot x_t$. From (5.3) and from conditions i) and ii) of Definition 5.3 we obtain for almost all $t \in [s', s]$

$$\begin{aligned} \dot{u}(t) &= -A((x_t, \dot{x}_t), u(t)) \\ &\geq -A((x_t, \dot{x}_t), \bar{v} \cdot x_t) + \rho \left[u(t) - \bar{v} \cdot x_t \right] \\ &> -M'(1 + |x_t|) + \underline{\rho} g(t). \end{aligned}$$

From this inequality and from condition iii) of Definition 5.1 we get

$$\dot{u}(t) - \bar{v} \cdot \dot{x}_t > -M'(1 + |x_t|) - |\bar{v}|(\alpha + \beta|x_t|) + \underline{\rho} g(t) > -M(1 + |x_t|) + \underline{\rho} g(t) \qquad \text{a.e..}$$

Using (5.1) and substituting the definition of $g(t)$ into the right hand side we obtain for almost every $t \in [s', s]$ the following:

$$
\begin{aligned}
\dot{u}(t) - \bar{v} \cdot \dot{x}_t \\
> -M \left[1 + \left(\frac{\alpha}{\beta} + |x_0| \right) e^{\beta t} - \frac{\alpha}{\beta} \right] \\
+ \frac{\rho M}{\rho - \beta} \left(1 + \frac{\alpha}{\beta} + |x_0| \right) \left[e^{\beta t} - e^{\rho t - (\rho - \beta)s} \right] \\
> -M \left(1 + \frac{\alpha}{\beta} + |x_0| \right) e^{\beta t} + \frac{\rho M}{\rho - \beta} \left(1 + \frac{\alpha}{\beta} + |x_0| \right) \left[e^{\beta t} - e^{\rho t - (\rho - \beta)s} \right] \\
= \frac{M}{\rho - \beta} \left(1 + \frac{\alpha}{\beta} + |x_0| \right) \left[\beta e^{\beta t} - \rho e^{\rho t - (\rho - \beta)s} \right] \\
= \dot{g}(t)
\end{aligned}
$$

The implication in (5.9) can be shown analogously. Finally, we observe that (5.8) and (5.9) imply the following: if $|u(t) - \bar{v} \cdot x_t| > g(t)$ for some $t \in [s', s]$, then $|u(t') - \bar{v} \cdot x_{t'}| > g(t')$ for all $t' \in [t, s]$. On the other hand, since $u(s) - \bar{v} \cdot x_s = g(s) = 0$ we must have $|u(t) - \bar{v} \cdot x_t| \leq g(t)$ for all $t \in [s', s]$. This completes the proof of the lemma. □

Lemma 5.4 *Let T be a β-bounded technology on the state space $X \subseteq \mathbf{R}^n$ and assume that $\beta > 0$. Furthermore, let $A : T \times \mathbf{R} \mapsto \mathbf{R}$ be an aggregator function for T and let $_0 x \in \mathbf{F}$ be a given feasible growth path. For every $s \in [0, \infty)$ there exists a unique absolutely continuous function $u^{(s)} : [0, s] \mapsto \mathbf{R}$ which satisfies the differential equation (5.3) and the boundary condition $u^{(s)}(s) = \bar{v} \cdot x_s$. Moreover, this function satisfies condition (5.7) on $[0, s]$.*

PROOF. According to the existence theorems for ordinary differential equations (see, e.g., Theorem 3 on p. 56 in [38]) there exists $\epsilon > 0$ and a unique absolutely continuous function $u^{(s)} : [s - \epsilon, s] \mapsto \mathbf{R}$ which solves (5.3) on the interval $[s - \epsilon, s]$ and which satisfies the boundary condition $u^{(s)}(s) = \bar{v} \cdot x_s$. Lemma 5.3 implies that $u^{(s)}$ remains uniformly bounded on its domain of definition. Therefore, it follows from familiar extension theorems for the solutions to ordinary differential equations that the function $u^{(s)}$ can be extended as a solution to (5.3) to the entire interval $[0, s]$. This completes the proof. □

We are now ready to prove the existence of a unique solution to Equation (5.3) which satisfies the growth condition (5.4).

Theorem 5.1 *Let T be a β-bounded technology on the state space $X \subseteq \mathbf{R}^n$ and assume that $\beta > 0$. Furthermore, let $A : T \times \mathbf{R} \mapsto \mathbf{R}$ be an aggregator* *function on T and let $_0x \in \mathbf{F}$ be a given feasible growth path. There exists a unique solution $u : [0,\infty) \mapsto \mathbf{R}$ to Equation (5.3) which satisfies condition (5.4). Moreover, we have*

$$|u(0)| \leq |\bar{v}||x_0| + K(x_0)$$

and

$$|u(0) - u^{(s)}(0)| \leq K(x_0)e^{-(\underline{\rho}-\beta)s} \tag{5.10}$$

for all $s \in [0,\infty)$. Here, the functions $u^{(s)} : [0,s] \mapsto \mathbf{R}$, $s \in [0,\infty)$, are defined as in Lemma 5.4.

PROOF. Because of (5.7) (evaluated at $t = 0$) we can find a sequence of positive real numbers $s_1, s_2, s_3, \ldots$ which diverges to infinity such that $\lim_{k\to\infty} u^{(s_k)}(0) = u_0$ exists and such that $|u_0| \leq |\bar{v}||x_0| + K(x_0)$ holds. Define $u : [0,\infty) \mapsto \mathbf{R}$ as a solution of Equation (5.3) with the initial condition $u(0) = u_0$. It is straightforward to verify that this solution exists on $[0,\infty)$ and that it satisfies $\lim_{k\to\infty} u^{(s_k)}(t) = u(t)$ for all $t \in [0,\infty)$. It follows from this property and from (5.7) that $|u(t)| \leq |\bar{v}||x_t| + K(x_0)e^{\beta t}$ holds. This shows that $|u(t)|e^{-\beta t}$ is uniformly bounded on $[0,\infty)$ by the constant $|\bar{v}|\|_0x\|_\beta + K(x_0)$ so that condition (5.4) is satisfied.

Now assume that there exists another solution u' of Equation (5.3) which satisfies (5.4). Without loss of generality we may assume $u'(0) > u(0)$. By the monotonicity of the aggregator function this implies $u'(t) > u(t)$ for all $t \in [0,\infty)$. Together with condition (5.5) this shows that

$$\dot{u}'(t) - \dot{u}(t) = A((x_t, \dot{x}_t), u(t)) - A((x_t, \dot{x}_t), u'(t)) \geq \underline{\rho}(u'(t) - u(t))$$

for almost every $t \in [0,\infty)$. Upon integration we obtain

$$u'(t) - u(t) \geq e^{\underline{\rho} t}(u'(0) - u(0)). \tag{5.11}$$

Because of $\rho > \beta$ this is a contradiction to our asumption that both u and u' satisfy (5.4). Thus, uniqueness of the solution u is proved.

Using the same argument as above one can also see that

$$|u^{(s')}(s) - u^{(s)}(s)| \geq e^{\underline{\rho} s}|u^{(s')}(0) - u^{(s)}(0)|$$

for all $0 < s < s'$. Because of $u^{(s)}(s) = \bar{v} \cdot x_s$ and because of (5.7) this inequality implies

$$|u^{(s')}(0) - u^{(s)}(0)| \leq K(x_0) \left[e^{-(\underline{\rho}-\beta)s} - e^{-(\underline{\rho}-\beta)s'} \right].$$

Choosing $s' = s_k$ we obtain the asserted inequality (5.10) as k approaches infinity. This completes the proof of the theorem. $\square$

Theorem 5.1 shows that we can define a utility functional $U : \mathbf{F} \mapsto \mathbf{R}$ which satisfies the recursive condition (5.2). It will be shown shortly that this functional is β-myopic, that is, continuous with respect to the β-topology. Furthermore, we shall prove that U is uniquely determined by Equation (5.2) within the class of all ϕ-bounded functionals, which is specified as follows. Define the functional $\phi : \mathbf{F}_\beta \mapsto \mathbf{R}$ by

$$\phi(_0 x) = 1 + |x_0|.$$

Since $|x_0| \leq \|_0 x\|_\beta$ it follows that ϕ is well defined on $\mathbf{F}_\beta$. A real-valued function f defined on $\mathbf{F}_\beta$ will be called ϕ-bounded, if

$$\sup \left\{ |f(_0 x)| / \phi(_0 x) \,\middle|\, _0 x \in \mathbf{F}_\beta \right\} < \infty.$$

We can now state the main result of this section.

Theorem 5.2 *For some $\beta > 0$ let T be a β-bounded technology on the state space $X \subseteq \mathbf{R}^n$ and let $A : T \times \mathbf{R} \mapsto \mathbf{R}$ be an aggregator function for T. For every feasible growth path $_0 x \in \mathbf{F}$ define the utility derived from this path by $U(_0 x) = u(0)$ where the function $u \in \mathcal{AC}(\mathbf{R})$ is the unique solution of Equation (5.3) which satisfies the growth condition (5.4) (see Theorem 5.1). Then it holds that $U : \mathbf{F} \mapsto \mathbf{R}$ is the unique ϕ-bounded functional satisfying condition (5.2). Moreover, U is continuous with respect to the β-topology on $\mathbf{F}$.*

PROOF. ϕ-boundedness of U follows immediately from $|u(0)| \leq |\bar{v}| |x_0| + K(x_0)$ (see Theorem 5.1) and from the definition of $K(x_0)$ in (5.6). Now assume that there is another functional $U' : \mathbf{F} \mapsto \mathbf{R}$ satisfying (5.2). Without loss of generality we may assume that $U'(_0 x) > U(_0 x)$ for some feasible growth path $_0 x$. By the same argument which led to (5.11) it follows that

$$U'(_t x) - U(_t x) \geq e^{\varrho t}(U'(_0 x) - U(_0 x))$$

for every $t \in [0, \infty)$. Because $|x_t| \leq e^{\beta t} \|_0 x\|_\beta$ we obtain

$$\frac{U'(_t x) - U(_t x)}{1 + |x_t|} \geq e^{(\varrho - \beta) t} \frac{U'(_0 x) - U(_0 x)}{e^{-\beta t} + \|_0 x\|_\beta} > e^{(\varrho - \beta) t} \frac{U'(_0 x) - U(_0 x)}{1 + \|_0 x\|_\beta}.$$

Since ϱ is strictly greater than β it follows that U' cannot be ϕ-bounded. Consequently, U is the only ϕ-bounded functional satisfying (5.2).

Finally, we have to prove continuity of U with respect to the β-topology. To this end we assume that $_0 x^{(k)}$ converges to $_0 \bar{x}$ in the β-topology as k approaches infinity and that $\epsilon > 0$ is an arbitrary positive real number. This implies that the sequence $x_0^{(k)}$ is bounded and, hence, that there exists a constant $\bar{K}$ such that $K(x_0^{(k)}) \leq \bar{K}$ holds for all $k \in \mathbf{Z}^+$. Now choose $s \in [0, \infty)$ so large that $\bar{K} e^{-(\varrho - \beta) s} < \epsilon/3$ and denote by $u^{(s,k)}$ the solution of the differential equation

$$\dot{u}(t) = -A((x_t^{(k)}, \dot{x}_t^{(k)}), u(t)) \qquad \text{a.e.}$$

which satisfies the boundary condition $u^{(s,k)}(s) = \bar{v} \cdot x_s^{(k)}$ (see Lemma 5.4). Analogously, define $\bar{u}^{(s)}$ as the solution of

$$\dot{u}(t) = -A((\bar{x}_t, \dot{\bar{x}}_t), u(t)) \qquad \text{a.e.}$$

satisfying $\bar{u}^{(s)}(s) = \bar{v} \cdot \bar{x}_s$. Now recall that convergence of $_0x^{(k)}$ to $_0\bar{x}$ implies uniform convergence on compact intervals as well as uniform convergence of the derivatives $_0\dot{x}^{(k)}$ almost everywhere on compact intervals. Because of the continuity of the aggregator function A this implies that $\lim_{k\to\infty} u^{(s,k)}(0) = \bar{u}^{(s)}(0)$ so that there exists a number $\bar{k} \in \mathbf{Z}^+$ such that $|u^{(s,k)}(0) - \bar{u}^{(s)}(0)| \leq \epsilon/3$ for all $k \geq \bar{k}$. Using these results as well as (5.10) we obtain for all $k \geq \bar{k}$ that

$$\begin{aligned}
|U(_0x^{(k)}) &- U(_0\bar{x})| \\
&\leq |U(_0x^{(k)}) - u^{(s,k)}(0)| + |u^{(s,k)}(0) - \bar{u}^{(s)}(0)| + |\bar{u}^{(s)}(0) - U(_0\bar{x})| \\
&\leq 2\bar{K}e^{-(\rho-\beta)s} + \frac{\epsilon}{3} \\
&< \epsilon.
\end{aligned}$$

Because $\epsilon > 0$ was chosen arbitrarily, it follows that U is continuous with respect to the β-topology. This completes the proof. $\qquad\square$

We would like to note that actually a slightly stronger continuity property holds, namely, that U is β'-myopic for all $\beta' \in (0, \rho)$. The proof of this result, which is an analogue to Lemma 2.6 in Part 1, requires some minor modifications in the proofs of the foregoing lemmas and theorems and will not be carried out here.

The following corollary is a simple consequence of the ϕ-boundedness of the utility functional U.

Corollary 5.1 *If the assumptions of Theorem 5.2 are satisfied, then there exists a constant $M > 0$ such that for every $x \in X$ and for every path $_0x \in \mathbf{F}(x)$ the inequality $|U(_0x)| \leq M(1 + |x|)$ holds.*

Most continuous time optimal growth models in the literature have been formulated with utility functionals in integral form, i.e.,

$$U(_0x) = \int_0^\infty e^{-\rho t} V(x_t, \dot{x}_t)\, dt$$

where $\rho > 0$ is a constant discount rate and $V : X \mapsto \mathbf{R}$ is a short-run utility function. This type of functionals represents the most important special case of recursive preferences. It can be obtained from Theorem 5.2 by specifying the aggregator function as

$$A((x, x'), z) = V(x, x') - \rho z. \qquad (5.12)$$

We conclude this section by presenting conditions on the aggregator function A which imply that the corresponding utility functional $U : \mathbf{F} \mapsto \mathbf{R}$ is concave on $\mathbf{F}$.

Lemma 5.5 *Let the assumptions of Theorem 5.2 be satisfied and assume, in addition, that the aggregator function $A : T \times \mathbf{R} \mapsto \mathbf{R}$ is concave. Then it holds that the utility functional $U : \mathbf{F} \mapsto \mathbf{R}$ is concave.*

PROOF. Let $_0x^{(0)}$ and $_0x^{(1)}$ be two feasible growth paths and let $u^{(s,0)}$ and $u^{(s,1)}$ be the corresponding solutions described in Lemma 5.4. Then we have $u^{(s,0)}(s) = \bar{v} \cdot x_s^{(0)}$ and $u^{(s,1)}(s) = \bar{v} \cdot x_s^{(1)}$ as well as

$$\dot{u}^{(s,0)}(t) = -A((x_t^{(0)}, \dot{x}_t^{(0)}), u^{(s,0)}(t)) \qquad \text{a.e.}$$

and

$$\dot{u}^{(s,1)}(t) = -A((x_t^{(1)}, \dot{x}_t^{(1)}), u^{(s,1)}(t)) \qquad \text{a.e..}$$

Defining $x_t^{(\lambda)} = (1 - \lambda)x_t^{(0)} + \lambda x_t^{(1)}$ and $u^{(s,\lambda)}(t) = (1 - \lambda)u^{(s,0)}(t) + \lambda u^{(s,1)}(t)$, it follows from the concavity assumptions that $u^{(s,\lambda)}(s) = \bar{v} \cdot x_s^{(\lambda)}$ and

$$\dot{u}^{(s,\lambda)}(t) \geq -A((x_t^{(\lambda)}, \dot{x}_t^{(\lambda)}), u^{(s,\lambda)}(t)) \qquad \text{a.e..}$$

From these properties and from the monotonicity assumption on the aggregator function with respect to its last argument we can conclude that every solution $u^{(s)}$ to the differential equation

$$\dot{u}(t) = -A((x_t^{(\lambda)}, \dot{x}_t^{(\lambda)}), u(t))$$

which satisfies $u^{(s)}(s) = \bar{v} \cdot x_s^{(\lambda)}$ must also satisfy $u^{(s)}(t) \geq u^{(s,\lambda)}(t)$ for all $t \in [0, s]$. Because of $\lim_{s \to \infty} u^{(s)}(0) = U(_0x^{(\lambda)})$ and

$$\lim_{s \to \infty} u^{(s,\lambda)}(0) = (1 - \lambda)U(_0x^{(0)}) + \lambda U(_0x^{(1)})$$

it follows immediately that U is concave. $\square$

As in the discrete time case it is possible to find aggregator functions which are not concave but which give rise to concave utility functionals. The conditions of Lemma 5.5 are therefore sufficient but not necessary for the concavity of U.

5.3 Optimal Growth Paths and Impatience

We are now in a position to present the definition of an optimal growth model which will be used throughout the remainder of this book.

Definition 5.4 Let $X \subseteq \mathbf{R}^n$ be a non-empty, closed, and convex set. An *optimal growth model* with state space X is a pair (T, A) where T is a β-bounded technology on X with $\beta > 0$ and $A : T \times \mathbf{R} \mapsto \mathbf{R}$ is an aggregator function for this technology. The optimal growth model is said to have *additively separable preferences* if there exists a constant $\rho > 0$ and a function $V : T \mapsto \mathbf{R}$ such that Equation (5.12) holds. The optimal growth model is said to be *concave* if the aggregator function A is concave.

As in Part 1 we define the *optimal value function* of the problem (T, A) by

$$W(x) = \sup \left\{ U({}_0x) \,\middle|\, {}_0x \in \mathbf{F}(x) \right\}. \tag{5.13}$$

In the present case, however, we have to take into account the fact that the feasible set $\mathbf{F}(x)$ can be empty for some state vectors $x \in X$. We handle this problem by adopting the convention that the supremum over the empty set is $-\infty$. Moreover, we extend the domain of definition of W to all of $\mathbf{R}^n$ by specifying $W(x) = -\infty$ for all $x \in \mathbf{R}^n \backslash X$. The optimal value function can therefore be considered as an extended-real-valued function on $\mathbf{R}^n$.

In accordance with viability theory (see [2] and [3]) we call a feasible state $x \in X$ *viable* if $\mathbf{F}(x)$ is non-empty. Furthermore, we denote by X^0 the set of all viable capital stocks, that is,

$$X^0 = \left\{ x \in X \,\middle|\, \mathbf{F}(x) \neq \emptyset \right\}.$$

It follows from Lemma 5.2 that X^0 is a convex subset of the state space X. If ${}_0x$ is a feasible growth path then it must hold that $x_t \in X^0$ for all $t \in [0, \infty)$. This shows that feasible state vectors which are not viable are irrelevant for all practical purposes and it is therefore convenient to define the set $T^0 \subseteq T$ by

$$T^0 = \left\{ (x, x') \in T \,\middle|\, x \in X^0 \right\}. \tag{5.14}$$

For viable state vectors $x \in X^0$ the supremum on the right hand side of (5.13) is always finite. This is a consequence of the following result which can be proved analogously to Lemma 2.8.[5]

Lemma 5.6 *The optimal value function W of an optimal growth model (T, A) is well defined (possibly equal to $-\infty$) and satisfies $|W(x)| \leq M(1 + |x|)$ for all $x \in X^0$. Here, $M > 0$ is a constant independent of x. If (T, A) is a concave optimal growth model, then W is a concave function and, hence, a continuous function on the relative interior of its effective domain $\operatorname{dom} W = X^0$.*

---

[5] The effective domain $\operatorname{dom} g$ of an extended-real-valued and concave function g is the set $\{x \,|\, g(x) > -\infty\}$. A concave function is known to be continuous on the relative interior of its effective domain (see [56]).

Definition 5.5 Let (T, A) be an optimal growth model with state space X, utility functional U, and optimal value function W. Moreover, let $x \in X$ be a feasible initial state such that $\mathbf{F}(x) \neq \emptyset$. An *optimal growth path* for the initial state x is a feasible growth path $_0x \in \mathbf{F}(x)$ such that $U(_0x) = W(x)$.

Let us spend a few words on the existence of optimal growth paths in the model (T, A). There is no way to adapt the simple argument used in the proof of Lemma 2.9 to the continuous time case even if we would take the existence of feasible paths for granted. This follows from the fact that the feasible set $\mathbf{F}(x)$ is in general not compact in any of the β-topologies, $\beta > 0$ (see the remark at the end of Section 5.1). Moreover, it is known from the theory of (time-additive) optimal control problems that optimal solutions do not necessarily exist without additional convexity assumptions (see, e.g., [20]). One would expect, however, that the existence proofs in [6] and [7] can be modified in such a way that they apply to the situation of a strictly concave optimal growth model in the sense of Definition 5.6. In particular, the compactness of the feasible sets $\mathbf{F}(x)$, $x \in X$, with respect to the compact open topology can be proved by exactly the same arguments as those used in [7, Lemma 6]. In a similar way one can show that the correspondence $x \mapsto \mathbf{F}(x)$ is closed and upper semi-continuous on dom W. The proof of the upper semi-continuity of the utility functionals U would perhaps require some new ideas and it is beyond the scope of this book to delve further into these matters. Suffice it to say that upper semi-continuity of the utility functionals would also imply upper semi-continuity of the optimal value function by the Maximum Theorem (see, e.g., [1, p. 67]).

As in the discrete time case we shall need a stronger assumption than concavity in order to prove the Minimum Impatience Theorems. Whereas for discrete time models it was sufficient to assume in addition to concavity of A that the mapping $x \mapsto A((x, x'), z)$ is strictly concave, we now require the strict concavity of the mapping $(x, x') \mapsto A((x, x'), z)$. One reason why we impose this stronger assumption is that without it the optimal policy functions of (T, A) need not be uniquely determined and continuous. It should be noted, however, that the assumptions formulated in the following definition are still weaker than strict concavity of $A((x, x'), z)$ with respect to all its arguments. The reader should also remember that strict monotonicity of the aggregator function with respect to future utility has already been postulated in Definition 5.3.

Definition 5.6 An optimal growth model (T, A) on the state space $X \subseteq \mathbf{R}^n$ is called *strictly concave* if the following condition is satisfied:

i) The aggregator function $A : T \times \mathbf{R} \mapsto \mathbf{R}$ is a concave function and the strict inequality

$$A\Big(((1-\lambda)x + \lambda\bar{x}, (1-\lambda)x' + \lambda\bar{x}'), (1-\lambda)z + \lambda\bar{z}\Big)$$
$$> (1-\lambda)A((x, x'), z) + \lambda A((\bar{x}, \bar{x}'), \bar{z})$$

holds for all $z \in \mathbf{R}$ and $\bar{z} \in \mathbf{R}$, for all $\lambda \in (0,1)$, and for all pairs $(x, x') \in T$ and $(\bar{x}, \bar{x}') \in T$ which satisfy $(x, x') \neq (\bar{x}, \bar{x}')$.

In the continuous time framework it is more convenient to work with the discount rate than with the discount factor. The Minimum Impatience Theorems will therefore be stated as lower bounds for the maximal discount rate along a given optimal growth path. To make this notion precise we introduce the following definition. For a motivation and some economic interpretations we refer back to Section 2.3.

Definition 5.7 Let (T, A) be an optimal growth model on the state space $X \subseteq \mathbf{R}^n$ and let $U : \mathbf{F} \mapsto \mathbf{R}$ be its utility function. Moreover, let $D \subseteq \mathbf{F}$ be an arbitrary non-empty set of feasible growth paths and define $\bar{\rho}(D)$ as the supremum over all $_0x \in D$, $t \in [0, \infty)$, and $\zeta \in \mathbf{R} \backslash \{0\}$ of the expression

$$\frac{A((x_t, \dot{x}_t), U(_tx) + \zeta) - A((x_t, \dot{x}_t), U(_tx))}{\zeta}.$$

We call $\bar{\rho}(D)$ the *maximal discount rate* of the aggregator function A on the set D. If $D = \mathbf{F}$, then we say that $\bar{\rho}(\mathbf{F})$ is the *globally maximal discount rate* of A.

Because of condition ii) of Definition 5.3 we have $\rho \leq \bar{\rho}(D)$ for all subsets $D \subseteq \mathbf{F}$. In case of the additively separable aggregator function defined in Equation (5.12) we have $\bar{\rho}(D) = \rho = \rho$ independently of the set D. In most of the results presented in Chapter 7 we shall specify D as a set consisting of a single growth path, $D = \{_0x\}$.

Chapter 6

Preliminary Results

In this chapter we summarize some results from dynamic programming and we show that the transformations introduced in Section 3.2 for the discrete time case can be applied equally well to continuous time optimal growth models. These results will be used in Chapter 7 to prove the Minimum Impatience Theorems.

In Section 6.1 we derive necessary and sufficient optimality conditions for the optimal growth models under consideration. Although these conditions resemble very closely their discrete time counterparts there is one crucial difference, namely, that they are stated only for concave models. The reason for this restriction are well known technical difficulties pertaining to the validity of the Bellman equation. More specifically, in the continuous time framework the Bellman equation can only be formulated in terms of the derivatives of the optimal value function and the assumption that these derivatives are well defined imposes implicit restrictions on the class of optimal growth models. We handle this difficulty by restricting ourselves for the most part to concave models, for which directional derivatives are known to exist. In view of the main purpose of this book we do not lose anything by such a restriction since the Minimum Impatience Theorems are only valid in strictly concave models anyway. A second difference as compared to the corresponding results in Section 3.1 is that we have to add interiority assumptions. This follows from the unboundedness of the subdifferential of a concave function on the boundary of its effective domain. The interiority assumptions will turn up again in the Minimum Impatience Theorems but it will be demonstrated in Chapter 7 that they do not substantially impede the applicability of those results. A third technical difficulty, although not a severe one, is the possible non-existence of feasible growth paths.

In Section 6.2 we present the continuous time counterparts of the concepts introduced in Section 3.2 for discrete time models, that is, equivalent optimal growth models, translations, and the time-τ transformation. Many proofs in this section are almost identical to those of the corresponding results in Section 3.2

and they will therefore be omitted.

There is no Section 6.3 on "Indeterminacy and Turnpikes" since the motivation for the Minimum Impatience Theorems is the same in the continuous time case as it was in the discrete time case. Let us just mention here that there exist continuous time versions of the Indeterminacy Theorem (see [48, 49, 61]) and of the Turnpike Theorem (see, e.g., [18, 19, 57, 60]). It is even more evident from continuous time Turnpike Theorems how the critical discount rate below of which convergence to a stationary growth path is guaranteed depends on the model under investigation. In [57], for example, the critical discount rate is explicitly related to the curvature properties of the Hamiltonian function of the problem. The other stability conditions have a similar structure and it is clear that they cannot be used to exclude the optimality of a given growth path in all strictly concave models on an appropriate state space. It is exactly this issue which is addressed and partly solved by the Minimum Impatience Theorems.

6.1 Dynamic Programming

The main purpose of this section is to prove a continuous time counterpart to Theorem 3.1, i.e., a necessary optimality condition based on the dynamic programming approach. Before we present this result, however, let us briefly consider Bellman's optimality principle. We have seen in Lemma 3.3 that this principle holds in discrete time optimal growth models provided that the aggregator function is strictly monotonic with respect to future utility. In continuous time models we have assumed strict monotonicity of the aggregator function from the very beginning (see Definition 5.3). Therefore, it comes at no surprise that the optimality principle holds without further assumptions. This result will be stated in Lemma 6.1 below. We omit the proof since it is an obvious modification of the proof of Lemma 3.3.

Definition 6.1 Let (T, A) be an optimal growth model with state space $X \subseteq \mathbf{R}^n$ and feasible set $\mathbf{F}$. We say that the *optimality principle* holds in this model if for all optimal paths $_0x \in \mathbf{F}$ and for all $t \in [0, \infty)$ the feasible path $_tx$ is also optimal.

Lemma 6.1 *Let (T, A) be an optimal growth model with state space $X \subseteq \mathbf{R}^n$. Then the optimality principle holds in (T, A).*

We have already indicated in the previous chapter that in contrast to discrete time models the feasible set $\mathbf{F}(x)$ can be empty for some initial capital stocks $x \in X$. This is not a big problem for our purpose because we are mainly interested in necessary optimality conditions which, by their very nature, are based on the explicit assumption that an optimal (and, henceforth, feasible) solution exists. The only consequence of the possible non-existence of feasible paths is that we have to restrict ourselves to the set of viable capital stocks X^0 which will be assumed to be non-empty throughout the remainder of the book.

Now let us turn to another technical issue which is more important than the possible non-existence of feasible growth paths. The Bellman equation for a continuous time dynamic programming problem is a partial differential equations and it can be a necessary optimality condition only if we make sure that the optimal value function W is sufficiently smooth. This is a real drawback since it is known that W can be a non-differentiable function even in very simple continuous time dynamic programming problems. Consequently, we have to sacrifice generality and restrict the class of optimal growth problems somehow. A reasonable way of doing this is to assume that the optimal growth models under consideration are concave. On the one hand, this causes actually no loss of generality for the main results of this book since we will have to make the concavity assumption for the Minimum Impatience Theorems anyway. On the other hand, concavity of the optimal growth model (T, A) implies concavity of the optimal value function W by Lemma 5.6 and, thereby, allows to draw from the powerful differential calculus for concave functions (see [56]). Some important properties of the optimal value function of a concave optimal growth model are summarized here for convenience. The optimal value function W is a concave (extended-real-valued) function on the space $\mathbf{R}^n$ and its effective domain is the set of viable states X^0. The subdifferential $\partial W(x)$ is a non-empty and bounded set for all $x \in \operatorname{int} X^0$. Moreover, the directional derivative

$$W'(x; x') = \lim_{\lambda \searrow 0} \frac{W(x + \lambda x') - W(x)}{\lambda}$$

is well defined (possibly $+\infty$ or $-\infty$) for all $x \in X^0$ and for all $x' \in \mathbf{R}^n$. The directional derivative $W'(x; x')$ and the subdifferential $\partial W(x)$ are related by the following identity which holds for all capital stocks x in the relative interior of the effective domain X^0:

$$W'(x; x') = \inf \left\{ x^* \cdot x' \,\middle|\, x^* \in \partial W(x) \right\}. \tag{6.1}$$

A very useful consequence of this relation is the following lemma.

Lemma 6.2 *Let $(s, s') \subseteq [0, \infty)$ be a non-empty interval and assume that the function $y : (s, s') \mapsto \mathbf{R}^n$ is absolutely continuous and satisfies $y(t) \in \operatorname{int} X^0$ for all $t \in (s, s')$. Then it follows that the function $t \mapsto W(y(t))$ is absolutely continuous on (s, s') and that the inequality*

$$\frac{d}{dt} W(y(t)) \geq W'(y(t); \dot{y}(t))$$

is satisfied for almost every $t \in (s, s')$.

PROOF. Because W is a concave function it is locally Lipschitz continuous on the interior of its effective domain X^0 (see, e.g., [55]). Given this property the absolute continuity of $t \mapsto W(y(t))$ follows as in [23, p. 158]. It is also shown there that

$$\frac{d}{dt}W(y(t)) \in \left\{ x^* \cdot \dot{y}(t) \,\middle|\, x^* \in \partial W(y(t)) \right\}$$

for almost every $t \in (s, s')$. The proof of the lemma is completed by combining this with Equation (6.1). $\square$

The optimal value function is finite and continuous on the relative interior of X^0 and it is infinite outside X^0. The lack of continuity on the boundary of the set of feasible initial states causes some directional derivatives to become infinite. This is the reason that the Bellman equation is a necessary optimality condition only on the interior of X^0 (see Theorem 6.1 below).

Before we state the main theorem of this section let us prove a preliminary result, namely, a functional inequality for the optimal value function. As a matter of fact, this inequality is already "half" of the Bellman equation.

Lemma 6.3 *Let (T, A) be a concave optimal growth model with state space $X \subseteq \mathbf{R}^n$. If W is the optimal value function of (T, A), then it holds that*

$$W'(x; x') + A((x, x'), W(x)) \leq 0 \tag{6.2}$$

for all $x \in \operatorname{int} X^0$ and for all $x' \in T_x$.

PROOF. Consider an arbitrary point $(\bar{x}, \bar{x}') \in T$ with $\bar{x} \in \operatorname{int} X^0$. From the properties of the set-valued mapping $x \mapsto T_x$ stated in Lemma 2.1 and from Michael's selection theorem (see [3, p. 82]) it follows that we can find a continuous selection f from T_x through the point $(\bar{x}, \bar{x}')$. This means that there exists a continuous function $f : X \mapsto \mathbf{R}^n$ with $f(\bar{x}) = \bar{x}'$ and $f(x) \in T_x$ for all $x \in X$. Consider the ordinary differential equation

$$\dot{y}(t) = f(y(t)). \tag{6.3}$$

Because f is continuous there exists a real number $\bar{\Delta} > 0$ and a solution $y : [0, \bar{\Delta}) \mapsto \mathbf{R}^n$ of (6.3) such that $y(0) = \bar{x}$. Furthermore, because $\bar{x} \in \operatorname{int} X^0$ one may choose $\bar{\Delta}$ so small that $y(\Delta) \in \operatorname{int} X^0$ for all $\Delta \in [0, \bar{\Delta})$. Note that by construction $(y(\Delta), \dot{y}(\Delta)) = (y(\Delta), f(y(\Delta))) \in T$ for all $\Delta \in [0, \bar{\Delta})$ so that y can be regarded as the initial segment of a feasible path in $\mathbf{F}(\bar{x})$. Moreover, $y(\Delta) \in \operatorname{int} X^0$ implies that $W(y(\Delta))$ is finite for all $\Delta \in [0, \bar{\Delta})$ so that we can find a feasible path $_0y^{(\Delta)} \in \mathbf{F}(y(\Delta))$ with utility $U(_0y^{(\Delta)}) \geq W(y(\Delta)) - \Delta^2$. Using this path we can define another path $_0x^{(\Delta)}$ as follows:

$$x_t^{(\Delta)} = \begin{cases} y(t) & \text{for } t \in [0, \Delta] \\ y_{t-\Delta}^{(\Delta)} & \text{for } t \in [\Delta, \infty) \end{cases}$$

Note that $_\Delta x^{(\Delta)} = {}_0 y^{(\Delta)}$ and that $_0 x^{(\Delta)} \in \mathbf{F}(\bar{x})$ provided that $\Delta \in [0, \bar{\Delta})$. The latter statement follows from the fact that $_0 x^{(\Delta)}$ is the concatenation of the feasible initial segment $y : [0, \Delta) \mapsto X^0$ and the feasible path $_0 y^{(\Delta)}$. From these properties we can conclude that

$$
\begin{aligned}
W(\bar{x}) &\geq U(_0 x^{(\Delta)}) \\
&= U(_0 x^{(\Delta)}) - U(_\Delta x^{(\Delta)}) + U(_0 y^{(\Delta)}) \\
&\geq -\int_0^\Delta \frac{d}{ds} U(_s x^{(\Delta)}) \, ds + W(y(\Delta)) - \Delta^2 \\
&= \int_0^\Delta A((y(s), \dot{y}(s)), U(_s x^{(\Delta)})) \, ds + W(y(\Delta)) - \Delta^2.
\end{aligned}
$$

The last line was obtained by using (5.2) and the definition of $_0 x^{(\Delta)}$. Because of the monotonicity of the aggregator function with respect to its last argument and because of $U(_s x^{(\Delta)}) \leq W(y(s))$ for all $s \in [0, \Delta]$ it follows from the above inequality upon rearranging terms and dividing by Δ that

$$
\Delta \geq \frac{W(y(\Delta)) - W(y(0))}{\Delta} + \frac{1}{\Delta} \int_0^\Delta A((y(s), \dot{y}(s)), W(y(s))) \, ds
$$

for all sufficiently small $\Delta > 0$. Since all functions involved in the integral are continuous on their relevant domains we can pass over to the limit as Δ approaches zero. In view of Lemma 6.2 this yields

$$
0 \geq \left. \frac{d}{dt} W(y(t)) \right|_{t=0} + A((y(0), \dot{y}(0)), W(y(0))) \geq W'(\bar{x}; \bar{x}') + A((\bar{x}, \bar{x}'), W(\bar{x})).
$$

Because $(\bar{x}, \bar{x}') \in T$ was chosen arbitrarily except for the condition $\bar{x} \in \text{int } X^0$ the proof of (6.2) is now complete. $\qquad\square$

We can dispense with the interiority condition in the above lemma if we happen to know that the optimal value function W is upper semi-continuous. Because this property would most likely be obtained as a by-product of a successful existence proof for optimal growth paths[1] we include a strengthened version of Lemma 6.3 under the upper semi-continuity assumption. It should be noted, however, that the other "half" of the Bellman equation (i.e., the $\geq$ sign in Inequality (6.2) along all optimal growth paths) does not hold without the interiority assumption even if the optimal value function is upper semi-continuous.

---

[1]We have mentioned this fact already in our discussion of the existence issue on page 119.

Lemma 6.4 *Let the conditions of Lemma 6.3 be satisfied and assume in addition that the interior of X^0 is non-empty and that the optimal value function W is upper semi-continuous. Then Inequality (6.2) holds for all $x \in X^0$ and for all $x' \in T_x$.*

PROOF. The inequality holds trivially if $W'(x; x') = -\infty$. In particular, this is true if $x + \lambda x' \notin X^0$ for all $\lambda > 0$. Therefore, we may assume without loss of generality that $W'(x; x') > -\infty$ and that there exists $\bar{\lambda} > 0$ such that $x + \lambda x' \in X^0$ for all $\lambda \in [0, \bar{\lambda}]$. Let μ be any real number satisfying $\mu < W'(x; x')$. From the definition of the directional derivative it follows that we can choose $\bar{\lambda} > 0$ so small that for all $\lambda \in (0, \bar{\lambda}]$ the following is true:

$$\frac{W(x + \lambda x') - W(x)}{\lambda} > \mu. \tag{6.4}$$

Let $(\bar{x}, \bar{x}')$ be an arbitrary point in the production technology set T such that $\bar{x} \in \text{int}\, X^0$ and define for all $k \in \mathbf{Z}^+$ the vectors[2]

$$x_k = \frac{k}{1+k} x + \frac{1}{1+k} \bar{x}$$

and

$$x'_k = \frac{k}{1+k} x' + \frac{1}{1+k} \bar{x}'.$$

From convexity of T it follows that $(x_k, x'_k) \in T$ for all $k \in \mathbf{Z}^+$. Moreover, we have $\lim_{k \to \infty} x_k = x$ and $\lim_{k \to \infty}(x_k + \lambda x'_k) = x + \lambda x'$. Now fix $\lambda \in (0, \bar{\lambda}]$ sufficiently small such that $\bar{x} + \lambda \bar{x}' \in \text{int}\, X^0$ (this is possible because $\bar{x} \in \text{int}\, X^0$). The upper semi-continuity of W and Corollary 7.5.1 in [56] imply that $\lim_{k \to \infty} W(x_k) = W(x)$ and $\lim_{k \to \infty} W(x_k + \lambda x'_k) = W(x + \lambda x')$. Together with (6.4) these properties show that for all sufficiently large $k \in \mathbf{Z}^+$ the following is true:

$$\frac{W(x_k + \lambda x'_k) - W(x_k)}{\lambda} > \mu.$$

Since the difference quotient in the above inequality is a non-increasing function of λ (see [56, Theorem 23.1]) we obtain $W'(x_k; x'_k) > \mu$. Together with the fact that $\mu < W'(x; x')$ was chosen arbitrarily this implies that

$$\liminf_{k \to \infty} W'(x_k; x'_k) \geq W'(x; x').$$

Finally, we observe that the inequality

$$W'(x_k; x'_k) + A((x_k, x'_k), W(x_k)) \leq 0$$

---

[2]The subscripts k do not represent a time argument.

holds for all $k \in \mathbf{Z}^+$ because of Lemma 6.3 and because of $x_k \in$ int X^0 for all $k \in \mathbf{Z}^+$. If we take into account all these properties and consider the inequality displayed last for $k \to \infty$ we obtain (6.2). This completes the proof. $\qquad \square$

After this short digression let us return to our main objective, namely, to the derivation of necessary optimality conditions for feasible growth paths in concave optimal growth models. We define the function $\psi : X \mapsto \mathbf{R}$ as in Section 3.1 by

$$\psi(x) = 1 + |x|.$$

Let $S \subseteq X$ be a non-empty set and let $f : S \mapsto \mathbf{R}$ be a function defined on S. We say that f is ψ-bounded on S whenever

$$\sup \left\{ |f(x)|/\psi(x) \,\Big|\, x \in S \right\} < \infty.$$

We can now prove the main result of this section.[3]

Theorem 6.1 *Let (T, A) be a concave optimal growth model with state space $X \subseteq \mathbf{R}^n$. There exist functions $G : T^0 \mapsto \mathbf{R} \cup \{-\infty, +\infty\}$ and $W : \mathbf{R}^n \mapsto \mathbf{R} \cup \{-\infty\}$ such that for every optimal growth path ${}_0x$ the following conditions are satisfied:*

i) *$G(x, x') \leq 0$ for all $x \in$ int X^0 and for all $x' \in T_x$. Moreover, $G(x_t, \dot{x}_t) = 0$ for almost all $t \in [0, \infty)$ for which $x_t \in$ int X^0.*

ii) *The function W is the optimal value function of (T, A) and, consequently, it is concave and ψ-bounded on X^0. Furthermore, the mapping $t \mapsto W(x_t)$ is absolutely continuous.*

iii) *The equation $A((x, x'), W(x)) = G(x, x') - W'(x; x')$ holds for all $(x, x') \in T^0$.*

If (T, A) is a strictly concave optimal growth model, then it holds that the function $x' \mapsto G(x, x')$ is strictly concave for all $x \in X^0$.

PROOF. Define W as the optimal value function of the model (T, A) and the function G as required by condition iii) of the theorem. Since non-negativity of $G(x, x')$ for all $x \in$ int X^0 and all $x' \in T_x$ follows from Lemma 6.3, condition i) of the theorem is verified if we can show that $G(x_t, \dot{x}_t) \geq 0$ holds for almost all t for which x_t is in the interior of X^0. Let us therefore consider an optimal growth path ${}_0x \in \mathbf{F}$. Because of Lemma 6.1 we know that for all $t \in [0, \infty)$ the

---

[3]Recall the definition of the set T^0 from Equation (5.14).

truncated path $_tx$ is also an optimal path. Therefore, we have $U(_tx) = W(x_t)$ and it follows from (5.2) that

$$\frac{d}{dt}W(x_t) = -A((x_t, \dot{x}_t), W(x_t)) \qquad \text{a.e..}$$

This implies in particular that the function $t \mapsto W(x_t)$ is absolutely continuous. Since concavity and ψ-boundedness of W follow from Lemma 5.6, we have already proved condition ii).

Now let $\Delta > 0$ be arbitrary and assume that $x_t \in \text{int } X^0$. Then we have

$$W(x_{t+\Delta}) - W(x_t) = \int_t^{t+\Delta} \frac{d}{ds}W(x_s)\,ds = -\int_t^{t+\Delta} A((x_s, \dot{x}_s), W(x_s))\,ds.$$

Because of this inequality and because of the concavity of W we obtain for every subgradient $x^* \in \partial W(x_t)$ the following:[4]

$$x^* \cdot \frac{x_{t+\Delta} - x_t}{\Delta} \geq -\frac{1}{\Delta}\int_t^{t+\Delta} A((x_s, \dot{x}_s), W(x_s))\,ds.$$

Taking into account the continuity of A and W (the latter on the interior of X^0) as well as the absolute continuity of the path $_0x$, we obtain as Δ approaches zero that

$$x^* \cdot \dot{x}_t + A((x_t, \dot{x}_t), W(x_t)) \geq 0 \qquad \text{a.e..}$$

Because this inequality holds for all $x^* \in \partial W(x_t)$ and because of (6.1) we finally obtain

$$W'(x_t; \dot{x}_t) + A((x_t, \dot{x}_t), W(x_t)) \geq 0 \qquad \text{a.e..}$$

This completes the proof of condition i) of the theorem. Condition iii) is satisfied by the construction of G.

If the model (T, A) is strictly concave, then we know that the function $A((x, x'), z)$ is strictly concave with respect to (x, x'). Together with the concavity of the directional derivative $W'(x; x')$ with respect to the variable x' (see [56, Theorem 23.1]) this implies that the function $x' \mapsto G(x, x')$ is strictly concave. The proof of the theorem is now complete. □

It should be noted that the interiority assumption in condition i) of the above theorem is essential if we insist on the property that the function W is the optimal value function. This is true even if the optimal value function is upper semi-continuous as can be seen from the following simple example with $n = 2$ and

$$X = \left\{(x, y) \,\middle|\, x^2 + y^2 \leq 1\right\},$$

$$T = \left\{(x, y, x', y') \,\middle|\, (x, y) \in X, (x', y') \in [-1, 1] \times [-1, 1]\right\},$$

$$A((x, y, x', y'), z) = -\rho z.$$

---

[4]The subdifferential $\partial W(x_t)$ is non-empty because of our assumption $x_t \in \text{int } X^0$.

Because the aggregator function does neither depend on the capital stocks nor on the investment rates we have $U \equiv 0$ and $W(x, y) = 0$ for all $(x, y) \in X$ and $W(x, y) = -\infty$ otherwise. Every feasible growth path is also optimal. Consider, for example, the path $(_0x, _0y)$ defined by

$$x_t = \sin t, \; y_t = \cos t.$$

This path extends along the boundary of $X = X^0$ for all $t \in [0, \infty)$. Obviously, it holds that $W'(x_t; \dot{x}_t) = -\infty$ for all $t \in [0, \infty)$ so that conditions i) and iii) of Theorem 6.1 cannot both be satisfied. If we did not insist on W being the optimal value function, then we could specify $W(x) = 0$ for all $x \in \mathbf{R}^2$ and the remaining conditions of Theorem 6.1 would be satisfied.

It follows from conditions i) and iii) of Theorem 6.1 that the Bellman equation

$$
\begin{aligned}
0 &= W'(x_t; \dot{x}_t) + A((x_t, \dot{x}_t), W(x_t)) \\
&= \sup \left\{ W'(x_t; x') + A((x_t, x'), W(x_t)) \,\Big|\, x' \in T_{x_t} \right\}
\end{aligned}
$$

holds for almost every $t \in [0, \infty)$ for which x_t is an element of the interior of X^0. A feasible growth path $_0x$ for which $x_t \in \operatorname{int} X^0$ holds for almost every $t \in [0, \infty)$ will be called an *almost interior* path. Note that this definition allows that the path $_0x$ has contact points at which it touches the boundary of X^0 but that it excludes non-trivial boundary segments. In contrast, we call a path $_0x$ an *interior* path if $x_t \in \operatorname{int} X^0$ for all $t \in [0, \infty)$. This precludes also the existence of contact points. Of course, every interior path is an almost interior path.

We proceed by proving that the necessary conditions of Theorem 6.1 are also sufficient for the optimality of a given almost interior growth path $_0x$.

Theorem 6.2 *Let (T, A) be a concave optimal growth model with state space $X \subseteq \mathbf{R}^n$ and let $_0x \in \mathbf{F}$ be an almost interior feasible growth path. If there exist functions $G : T^0 \mapsto \mathbf{R} \cup \{-\infty, +\infty\}$ and $W : \mathbf{R}^n \mapsto \mathbf{R} \cup \{-\infty\}$ such that the conditions i) - iii) of Theorem 6.1 are satisfied, then it follows that $_0x$ is an optimal growth path for the model (T, A).*

PROOF. First let us consider a fixed number $s \in [0, \infty)$ for which $x_s \in \operatorname{int} X^0$. Because $_0x$ is a continuous function it holds that $x_t \in \operatorname{int} X^0$ for all t in a sufficiently small neighborhood V of s. From concavity and ψ-boundedness of W it follows that W is subdifferentiable at x_t for all t in V. Now consider the function $t \mapsto W(x_t)$. From Lemma 6.2 we know that it is absolutely continuous and that

$$\frac{d}{dt} W(x_t) \geq W'(x_t; \dot{x}_t)$$

holds for almost all t in the given neighborhood V. Using this fact we obtain from the condition $G(x_t, \dot{x}_t) = 0$ that

$$\frac{d}{dt} W(x_t) \geq -A((x_t, \dot{x}_t), W(x_t))$$

almost everywhere on V. Together with $W(x_t) \geq U({}_tx)$, Inequality (5.5), and Equation (5.2) this implies that

$$\frac{d}{dt} W(x_t) \geq -A((x_t, \dot{x}_t), U({}_tx)) + \underline{\rho}(W(x_t) - U({}_tx))$$

$$= \frac{d}{dt} U({}_tx) + \underline{\rho}(W(x_t) - U({}_tx))$$

almost everywhere on V.

So far, t has been restricted to the neighborhood V of a particular point s for which x_s is in the interior of X^0. Because ${}_0x$ is assumed to be an almost interior path, however, we can conclude that the above relations hold for almost all $t \in [0, \infty)$. Because the function $t \mapsto W(x_t)$ is absolutely continuous by condition ii) of Theorem 3.1 we can integrate the inequality derived last to obtain

$$W(x_t) - U({}_tx) \geq e^{\underline{\rho} t}(W(x_0) - U({}_0x)) \tag{6.5}$$

for all $t \in [0, \infty)$. From Corollary 5.1 and from Formula (5.1) it follows that there exists a constant $M > 0$ such that

$$|U({}_tx)| \leq M(1 + |x_t|) \leq M\left[1 + \left(\frac{\alpha}{\beta} + |x_0|\right)e^{\beta t} - \frac{\alpha}{\beta}\right]$$

for all $t \in [0, \infty)$. Substituting this into (6.5) we obtain

$$W(x_t) \geq e^{\underline{\rho} t}(W(x_0) - U({}_0x)) - M\left[1 + \left(\frac{\alpha}{\beta} + |x_0|\right)e^{\beta t} - \frac{\alpha}{\beta}\right].$$

Now assume that $W(x_0) - U({}_0x) > 0$. It follows from $\underline{\rho} > \beta$ that the first term on the right hand side of the above inequality dominates the second term as t approaches infinity. Consequently, $W(x_t)$ increases exponentially with rate $\underline{\rho}$. Employing a by now familiar argument based on the β-boundedness of the technology T we see that W cannot be ψ-bounded. This is a contradiction to condition ii) of Theorem 6.1 so that $U({}_0x) = W(x_0)$ must hold. This, in turn, shows that ${}_0x$ is an optimal path and the proof of the theorem is complete. $\square$

Let us combine the results derived so far to prove the Bellman equation for strictly concave optimal growth models and to state the necessary optimality conditions in terms of an optimal policy function h which determines the optimal growth paths as the trajectories of the dynamical system

$$\dot{x}_t = h(x_t). \tag{6.6}$$

To be more specific, we define an optimal policy function in the following way.

Definition 6.2 Let (T, A) be an optimal growth model with state space $X \subseteq \mathbf{R}^n$ and let $S \subseteq X$ be an arbitrary subset of X. A function $h : S \mapsto \mathbf{R}^n$ is called an *optimal policy function* for (T, A) if for all initial states $x_0 \in S$ it holds that Equation (6.6) has a unique solution $x_0 \in \mathcal{AC}(S)$ and that this solution $_0x$ is an optimal growth path for (T, A).

It is trivial to see that the set S in the above definition must be a subset of the set X^0 of viable capital stocks of (T, A). The reader might be wondering why we have included the set S at all and why we did not simply define optimal policy functions on $S = X^0$. The reason for the distinction between the domain S of h and the set of viable initial states X^0 will become apparent when we formulate the Minimum Impatience Theorem in terms of a given policy function h. Briefly stated, the argument runs as follows: since our goal is to exclude the optimality of h in any strictly concave optimal growth model, we do not want to specify the set of viable initial states X^0 in advance, because this would already restrict the class of optimal growth models under consideration.

Theorem 6.3 *Let (T, A) be a strictly concave optimal growth model with state space $X \subseteq \mathbf{R}^n$ and let $X^0 \subseteq X$ be the set of viable states. Assume that $S \subseteq X^0$ is a given set and $h : S \mapsto \mathbf{R}^n$ is an optimal policy function. Then it follows that the optimal value function W of (T, A) is strictly concave and satisfies*

$$\sup \left\{ W'(x; x') + A((x, x'), W(x)) \,\middle|\, x' \in T_x \right\} = 0 \tag{6.7}$$

for all $x \in S \cap \operatorname{int} X^0$. Moreover, it holds that the supremum on the left hand side of Equation (6.7) is attained uniquely at the point $x' = h(x)$ and that the function h is continuous on the set $S \cap \operatorname{int} X^0$.

PROOF. If h is an optimal policy function, then it follows from Theorem 6.1 that

$$W'(x; h(x)) + A((x, h(x)), W(x)) = 0$$

holds for all $x \in S \cap \operatorname{int} X^0$. In view of Lemma 6.3 this implies that Equation (6.7) is satisfied for all $x \in S \cap \operatorname{int} X^0$. Since the function $x' \mapsto W(x; x') + A((x, x'), W(x))$ is strictly concave by Theorem 6.1 it follows that the supremum in (6.7) is attained only at $x' = h(x)$. Finally, for every $x \in \operatorname{int} X^0$ it holds that the directional derivative $W'(x; x')$ is a finite and concave function with respect to $x' \in \mathbf{R}^n$. Because this implies continuity of the map $x' \mapsto W'(x; x') + A((x, x'), W(x))$ the continuity of h follows from Lemma 5.1 and the Maximum Theorem. This completes the proof. □

6.2 Transformations

In this section we summarize some results on equivalent optimal growth models, translations, and the time-τ transformation which will be needed in Chapter 7. Since these results are very similar to their discrete time counterparts, we shall be rather brief and state most of the lemmas without proofs.

Definition 6.3 Let (T, A) and (T, A') be two optimal growth models with a common technology T defined on the state space $X \subseteq \mathbf{R}^n$ and, hence, a common set of feasible growth paths $\mathbf{F}$. The two models are said to be *equivalent* if the following property is true: a feasible growth path $_0x \in \mathbf{F}$ is an optimal path for the model (T, A) if and only if it is also an optimal path for the model (T, A').

In Section 3.2 we have introduced *translations* as a method of generating an optimal growth model (T, A') which is equivalent to a given model (T, A). This method is also applicable in the continuous time framework when one defines the translation formula in the following way:

$$A'((x, x'), z) = A((x, x'), z - p \cdot x) - p \cdot x'. \tag{6.8}$$

The vector $p \in \mathbf{R}^n$ is called the *translation vector*. The following lemma can be proved in almost exactly the same way as Lemma 3.6.

Lemma 6.5 *Let (T, A) be an optimal growth model with state space $X \subseteq \mathbf{R}^n$ and let U be the utility functional generated by the aggregator function A. Moreover, let $p \in \mathbf{R}^n$ be an arbitrary vector and define the function $A' : T \times \mathbf{R} \mapsto \mathbf{R}$ by (6.8). Then it holds that A' is an aggregator function for T, that A' generates the utility functional $U'(_0x) = U(_0x) + p \cdot x_0$, and that the models (T, A) and (T, A') are equivalent.*

Properties of an optimal growth model which are preserved by translations will be called *translation invariant*. Some of these properties are listed in the following lemma. The proof is very simple and will not be presented here.

Lemma 6.6 *Let the assumptions of Lemma 6.5 be satisfied. Furthermore, assume that $D \subseteq \mathbf{F}$ is a non-empty set and $\rho > 0$ a real number. The following properties are translation invariant, i.e., they are valid in the model (T, A) if and only if they are valid in the model (T, A'):*

i) The optimal growth model is additively separable.

ii) The optimal growth model is (strictly) concave.

iii) The aggregator function satisfies condition (5.5) with $\underline{\rho} = \rho$.

iv) The maximal discount rate of the aggregator function on the set D is equal to ρ.

As in the discrete time case we can use translations to generate equivalent optimal growth models satisfying certain monotonicity conditions.

Lemma 6.7 *Let (T, A) be an optimal growth model with state space $X \subseteq \mathbf{R}^n$ and assume that the function $(x, x') \mapsto A((x, x'), z)$ from T to $\mathbf{R}$ satisfies a Lipschitz condition with a Lipschitz constant M which is independent of z. Then there exists a vector $p \in \mathbf{R}^n$ such that the optimal growth model (T, A') obtained from (T, A) by a translation along the vector p has the following properties:*

i) The function $x \mapsto A'((x, x'), z)$ from $T_{x'} = \{x \in X \,|\, (x, x') \in T\}$ to $\mathbf{R}$ is strictly increasing for all $z \in \mathbf{R}$ and for all $x' \in X$ for which $T_{x'}$ is non-empty.

ii) The function $x' \mapsto A'((x, x'), z)$ from T_x to $\mathbf{R}$ is strictly decreasing for all $x \in X$ and for all $z \in \mathbf{R}$.

PROOF. Define the vector $p \in \mathbf{R}^n$ by $p = (k, k, \ldots, k)$ where k is any real number satisfying $k > \max\{\rho^{-1}M, M\}$ and proceed as in Lemma 3.8. We omit the details. $\square$

The proof of the following result is identical to the one of Lemma 3.9 and it will not be repeated here.

Lemma 6.8 *Let (T, A) be a concave optimal growth model with state space $X \subseteq \mathbf{R}^n$ and optimal value function $W : X \mapsto \mathbf{R}$. Let $Y \subseteq \mathbf{R}^n$ be an affine set such that $X^0 \cap Y \neq \emptyset$ where $X^0 \subseteq X$ is the set of viable initial states. Moreover, denote the restriction of the function W to the set $X^0 \cap Y$ by W_Y. If $y_0 \in X^0 \cap Y$ is a point at which the concave function W_Y is subdifferentiable, then there exists a vector $p \in \mathbf{R}^n$ such that the concave optimal growth model (T, A') obtained from (T, A) by a translation along the vector p has the following property: the optimal value function W' of (T, A') attains its maximum on the set $X^0 \cap Y$ at the given point y_0.*

The time-τ transformation for discrete time optimal growth models was introduced in Section 3.2 as a means to transform a given model (T, A) in such a way that the τ-th iterate of an optimal policy function of this model becomes an optimal policy function of the transformed model (T', A'). A similar transformation is also possible in the continuous time framework as will be demonstrated in the remainder of this section. Since there is no conceptual difference to the case of discrete time models we shall omit some details of the proof.

Let us assume that (T, A) is a given optimal growth model with state space $X \subseteq \mathbf{R}^n$ and let $h : S \mapsto \mathbf{R}^n$ be an optimal policy function. Here we assume that T is a β-bounded technology and that $S \subseteq X^0$ is a non-empty, closed, and convex set. From Definition 6.2 it follows that for every $x \in S$ there exists a unique trajectory $_0x \in \mathcal{AC}(S)$ such that $x_0 = x$ and

$$\dot{x}_t = h(x_t) \tag{6.9}$$

for almost all $t \in [0, \infty)$. In a slight abuse of notation we denote by $h^{(\tau)}(x)$ the value of this trajectory at time τ, i.e., $h^{(\tau)}(x) = x_\tau$. Note that $h^{(\tau)}$ is not an iterate of h but the so-called time-τ map associated with the differential equation (6.9). The time-τ map is a continuous mapping from S to S and it satisfies the semi-group properties $h^{(0)} = \mathrm{id}_S$ and $h^{(\tau+\tau')} = h^{(\tau)} \circ h^{(\tau')}$ for all $\tau, \tau' \in [0, \infty)$.[5] We want to show that there exists a discrete time optimal growth model (T', A') with state space S such that $h^{(\tau)}$ is an optimal policy function for this model.

To define the technology T' we proceed as in Section 3.2. We say that a vector $x' \in S$ is τ-reachable from another vector $x \in S$ if there exists a feasible growth path $_0x \in \mathbf{F}(x)$ such that $x' = x_\tau$. Now we can define

$$T' = \Big\{ (x, x') \,\Big|\, x \in S \text{ and } x' \text{ is } \tau\text{-reachable from } x \Big\}.$$

It is straightforward to verify that T' is a technology on the state space S in the sense of Definition 2.1. Moreover, if T is β-bounded, then T' is β'-bounded with $\beta' = e^{\beta\tau}$.

The aggregator function $A' : T' \times \mathbf{R} \mapsto \mathbf{R}$ is defined by means of a finite horizon optimal control problem. More specifically, we define for all $(x, x') \in T'$ and for all $z \in \mathbf{R}$

$$A'((x, x'), z) = \sup u(0)$$

where the supremum is taken over the set of all functions $u \in \mathcal{AC}(\mathbf{R})$ and $y \in \mathcal{AC}(S)$ subject to the constraints

$$\dot{u}(t) = -A((y(t), v(t)), u(t)) \qquad \text{a.e. on } [0, \tau],$$
$$\dot{y}(t) = v(t) \qquad \text{a.e. on } [0, \tau],$$
$$u(\tau) = z,$$
$$y(0) = x,$$
$$y(\tau) = x',$$
$$(y(t), v(t)) \in T \qquad \text{a.e. on } [0, \tau].$$

---

[5] At this point the reader will probably appreciate the notation $h^{(\tau)}$ because the time-τ map is obviously the proper generalization of the τ-th iterate of a mapping to a continuous flow defined by a differential equation.

We denote this control problem by $P(x, x', z)$ as it depends on the parameters x, x', and z. The variables u and y represent the state variables of $P(x, x', z)$ and the variable v is the control variable. Note that $A'((x, x'), z)$ is well defined for all $(x, x') \in T'$ and all $z \in \mathbf{R}$ because the problem $P(x, x', z)$ has at least one feasible solution by the definition of T'.

The next step is to verify that the aggregator function A' defined in this way satisfies the Lipschitz condition (2.5). To this end assume that u and u' are admissible state trajectories for the problems $P(x, x', z)$ and $P(x, x', z')$, respectively. In the same way as in Section 5.2 (see Inequality (5.11)) we obtain

$$|u'(t) - u(t)| \geq e^{\varrho t}|u'(0) - u(0)|$$

for all $t \in [0, \tau]$. In particular, for $t = \tau$ we get

$$|u'(0) - u(0)| \leq e^{-\varrho \tau}|z' - z|$$

which shows that A' satisfies the Lipschitz condition (2.5) with $\bar{\delta} = e^{-\varrho \tau}$. The other conditions of Definition 2.3 can be verified in a similar way and we conclude that (T', A') is indeed a discrete time optimal growth model. Moreover, it is not difficult to show that (T', A') is additively separable, concave, or strictly concave, whenever the original model (T, A) has the corresponding property. We would also like to mention that the globally minimal discount factor of the transformed aggregator function, $\underline{\delta}(\mathbf{F}')$, is greater than or equal to the quantity $e^{-\tau \bar{\rho}(\mathbf{F})}$ where $\bar{\rho}(\mathbf{F})$ denotes the globally maximal discount rate of the original aggregator A.

Finally, we claim that the restriction to S of the optimal value function W of the model (T, A) is the optimal value function W' of the transformed model (T', A'), that is, $W'(x) = W(x)$ for all $x \in S$. To this end we first observe that

$$W(x) = \sup \left\{ U(_0 x) \,\middle|\, _0 x \in \mathbf{F}(x) \right\}$$
$$= \sup_{x' \in T'_x} \sup \left\{ U(_0 x) \,\middle|\, _0 x \in \mathbf{F}(x), x_\tau = x' \right\}. \tag{6.10}$$

Moreover, we have for every fixed pair $(x, x') \in T'$

$$\sup \left\{ U(_0 x) \,\middle|\, _0 x \in \mathbf{F}(x), x_\tau = x' \right\}$$
$$= \sup \left\{ -\int_0^\tau \frac{d}{dt} U(_t x) \, dt + U(_\tau x) \,\middle|\, _0 x \in \mathbf{F}(x), x_\tau = x' \right\}$$
$$= \sup \left\{ \int_0^\tau A((x_t, \dot{x}_t), U(_t x)) \, dt + W(x') \,\middle|\, _0 x \in \mathbf{F}(x), x_\tau = x' \right\}$$
$$= \sup \left\{ u(0) \,\middle|\, \dot{u}(t) = -A((x_t, \dot{x}_t), u(t)), u(\tau) = W(x'), _0 x \in \mathbf{F}(x), x_\tau = x' \right\}.$$

The expression in the last line is, by definition, equal to $A'((x, x'), W(x'))$ so that we get

$$\sup\left\{U(_0x) \,\middle|\, _0x \in \mathbf{F}(x),\, x_\tau = x'\right\} = A'((x, x'), W(x')).$$

Substituting this into (6.10) we finally obtain

$$W(x) = \sup\left\{A'((x, x'), W(x')) \,\middle|\, x' \in T'_x\right\}$$

which shows that W solves the Bellman equation of the problem (T', A'). In view of Lemma 3.1 our claim is therefore proved. Moreover, it is obvious from the above derivations that the supremum in the Bellman equation is attained at $x' = h^{(\tau)}(x)$.

For later reference we summarize these result in the following lemma.

Lemma 6.9 *Let (T, A) be a (continuous time) optimal growth model on the state space $X \subseteq \mathbf{R}^n$ and let $h : S \mapsto \mathbf{R}^n$ be an optimal policy function for this model. Assume that $S \subseteq X$ is non-empty, closed, and convex. For every $\tau > 0$ there exists a discrete time optimal growth model (T', A') on the state space S which has the time-τ map $h^{(\tau)}$ as its optimal policy function. Moreover, the following properties are satisfied:*

  i) If T is a β-bounded technology, then T' is a $e^{\beta\tau}$-bounded technology.

  ii) The globally minimal discount factor of A' is greater than or equal to $e^{-\tau\bar{\rho}(\mathbf{F})}$ where $\bar{\rho}(\mathbf{F})$ denotes the globally maximal discount rate of A.

  iii) The restriction to S of the optimal value function of (T, A) is the optimal value function of (T', A').

  iv) If (T, A) is concave (strictly concave, additively separable), then the same holds true for the transformed model (T', A').

Chapter 7

Minimum Impatience Theorems

In this chapter we present the central results of Part 2, namely, the Minimum Impatience Theorems and their various corollaries for continuous time optimal growth models. The basic ideas of the proofs in this chapter are the same as in the discrete time case although the technical details differ a little bit.

In Section 7.1 we present the main theorems and we discuss the way how they can be applied. The first theorem, a continuous time counterpart to Theorem 4.1, provides a lower bound on the maximal discount rate along a given growth path in any strictly concave optimal growth model for which the given path is an almost interior optimal solution. The restriction to almost interior paths is new as compared to the discrete time case and it is a consequence of the corresponding interiority assumptions in the dynamic programming conditions of Section 6.1. Fortunately, this additional assumption does not cause a severe restriction on the possible applications of our results. As a matter of fact, in many interesting cases it is possible to conclude directly from the definition of a feasible path that it can only be an almost interior one. We shall explain this issue further and illustrate it by an example in Section 7.1. This type of Minimum Impatience Theorem will be stated both for a single almost interior growth path and for a policy function h. To avoid any difficulties pertaining to paths which are not almost interior in the latter case, we assume that h is defined on an open set $S \subseteq X$. Since we know that S must be a subset of the set of viable initial states X^0 (but does not have to coincide with it), every path generated by h is necessarily an interior path.

It turns out that the Minimum Impatience Theorems mentioned in the preceding paragraph are conclusive only for optimal growth paths which exhibit a high degree of non-monotonicity. They are, therefore, of no use in one-dimensional continuous time optimal growth models. This follows from the fact that one-dimensional differential equations can only have monotonic solutions. In order to illustrate our results we have therefore chosen a two-dimensional system of differential equations. It is one of the simplest examples which exhibits a limit cycle as a possible solution trajectory. Monotonic optimal growth paths (and, in particular, policy functions for one-dimensional problems) can be dealt with by another type of Minimum Impatience Theorems. The continuous time analogue to Theorem 4.4 is of this type and will be derived at the end of Section 7.1.

For a given policy function we can construct the associated average correspondence in a similar way as this was done in Chapter 4 for the discrete time case. We discuss some basic properties and approximation results for the average correspondence in Section 7.2. Finally, in Section 7.3 we use these approximation results to derive Minimum Impatience Theorems which are based on local properties of the given policy function. More specifically, it is enough to know the Jacobian matrix of the policy function at an interior stationary state. If this matrix is ρ-expanding (a term to be defined in Section 7.3) or if it has a real eigenvalue λ with multiplicity one and $\lambda \geq \rho$, then we can conclude that ρ is a lower bound on the maximal rate of impatience required for the optimality of the given policy function. In contrast to the discrete time framework only the second local Minimum Impatience Theorem (the one based on the existence of a real eigenvalue λ with $\lambda \geq \rho$) can be generalized to the case where the policy function has a periodic trajectory of arbitrary period τ. As a matter of fact, it turns out that the auxiliary problem derived from the original problem by an application of the time-τ transformation has an optimal stationary state $\bar{x}$ instead of the given periodic path but that this stationary state $\bar{x}$ cannot be δ-expanding for any δ smaller than one.[1]

7.1 Main Results

In this section we prove the Minimum Impatience Theorems for strictly concave optimal growth models in continuous time. As in the discrete time case we first present a result where we assume that a single growth path is given. The situation in which more than one path or a complete policy function h is given will be discussed later.

Let us assume that $X \subseteq \mathbf{R}^n$ is a non-empty, closed, and convex set and that $_0x \in \mathcal{AC}(X)$ is an arbitrary function. For every $\rho \in [0, \infty)$ we denote by $\tilde{H}(_0x, \rho)$ the set of all vectors $y \in X$ for which there exists a piecewise absolutely continuous function $\mu : [0, \infty) \mapsto \mathbf{R}$ such that the following conditions are

---

[1] The auxiliary problem is a discrete time problem so that δ-expansiveness for some $\delta < 1$ would be necessary to derive a lower bound on the time-preference rate.

satisfied:[2]

$$\mu : [0, \infty) \mapsto \mathbf{R} \text{ is non-negative and non-increasing} \tag{7.1}$$

$$\int_0^\infty e^{-\rho t} \mu(t) \, dt = 1 \tag{7.2}$$

$$\int_0^\infty e^{-\rho t} \mu(t) x_t \, dt = y \tag{7.3}$$

Note that the function μ occurring in conditions (7.1) - (7.3) need not be continuous. Before we state the main result let us show that the correspondence $\rho \mapsto \tilde{H}(_0x, \rho)$ is non-increasing.

Lemma 7.1 *Let $_0x \in \mathcal{AC}(X)$ be a given path and let ρ and ρ' be two real numbers such that $0 \le \rho' \le \rho$. Then it holds that $\tilde{H}(_0x, \rho) \subseteq \tilde{H}(_0x, \rho')$.*

PROOF. Assume that $y \in \tilde{H}(_0x, \rho)$. Then there must exist a piecewise absolutely continuous function $\mu : [0, \infty) \mapsto \mathbf{R}$ such that conditions (7.1) - (7.3) are satisfied. Defining $\mu'(t) = \mu(t)e^{(\rho'-\rho)t}$ for all $t \in [0, \infty)$ it can easily be verified that these conditions remain true when ρ and $\mu(t)$ are replaced by ρ' and $\mu'(t)$, respectively. Moreover, the function μ' is piecewise absolutely continuous. Therefore, it follows that $y \in \tilde{H}(_0x, \rho')$ and the proof is complete. $\square$

We can now state and prove the first version of a continuous time Minimum Impatience Theorem.

Theorem 7.1 *Let $X \subseteq \mathbf{R}^n$ be a non-empty, closed, and convex set and let $_0x \in \mathcal{AC}(X)$ be an arbitrary function. Assume that $_0x$ is not constant on any interval of the form $[0, \epsilon)$ with $\epsilon > 0$ and define the set D by $D = \{_0x\}$. If $_0x$ is an almost interior optimal growth path in a strictly concave optimal growth model (T, A) with viable initial states $X^0 \subseteq X$ and if $x_0 \in \text{int } X^0$, then it holds that $x_0 \notin \tilde{H}(_0x, \rho)$ for all $\rho \ge \bar{\rho}(D)$. Here $\bar{\rho}(D)$ denotes the maximal discount rate of A along the path $_0x$.*

---

[2] The function μ is said to be piecewise absolutely continuous if there exists a sequence of numbers t_k, $k \in \mathbf{Z}_0^+$, such that the following conditions are satisfied:

i) The set $\{k \in \mathbf{Z}_0^+ \,|\, t_k \in [0, M]\}$ is finite for all $M \in [0, \infty)$ and it holds that $0 = t_0 < t_1 < t_2 < \dots$.

ii) μ is absolutely continuous on each of the intervals (t_k, t_{k+1}) and μ is continuous from the right on all of $[0, \infty)$.

iii) The possible discontinuities of μ at the points t_k, $k \in \mathbf{Z}^+$, are jumps of finite height.

PROOF. Because of Lemma 7.1 it is sufficient to show that $x_0 \notin \tilde{H}(_0 x, \rho)$ for $\rho = \bar{\rho}(D)$. The proof is by contradiction. Assuming that $x_0 \in \tilde{H}(_0 x, \rho)$ it follows that there exists a piecewise absolutely continuous function $\mu : [0, \infty) \mapsto \mathbf{R}$ such that conditions (7.1) and (7.2) hold and such that

$$\int_0^\infty e^{-\rho t} \mu(t) x_t \, dt = x_0. \tag{7.4}$$

Since $_0 x$ is feasible in an optimal growth model (T, A), there exist constants $\beta > 0$ and $\underline{\rho} \in [0, \infty)$ such that the conditions $_0 x \in \mathbf{F}_\beta$ and $\rho = \bar{\rho}(D) \geq \underline{\rho} > \beta$ are satisfied. Defining $\lambda(t) = e^{-\rho t} \mu(t)$ for all $t \in [0, \infty)$ it follows that $\lambda(t) \geq 0$ for all $t \in [0, \infty)$ as well as

$$\int_0^\infty \lambda(t) \, dt = 1 \tag{7.5}$$

and

$$\int_0^\infty \lambda(t) x_t \, dt = x_0. \tag{7.6}$$

Now let us consider the integral

$$\int_0^\infty \lambda(t) \dot{x}_t \, dt.$$

Because of $0 \leq \lambda(t) \leq \mu(0) e^{-\rho t} \leq \mu(0) e^{-\underline{\rho} t}$, $_0 x \in \mathbf{F}_\beta$, and $\beta < \underline{\rho}$ it follows that this infinite integral is absolutely convergent. This shows that the vector

$$y_0 = \int_0^\infty \lambda(t) \dot{x}_t \, dt \tag{7.7}$$

is well defined. The assumption that x_0 is an interior point of the set X^0 of viable initial states of (T, A) implies that the optimal value function of (T, A) is subdifferentiable at x_0. In view of Lemma 6.8 and Lemma 6.6 we may therefore assume that the optimal value function W of (T, A) satisfies

$$W(x_0) \geq W(x_t) \tag{7.8}$$

for all $t \in [0, \infty)$. Equations (7.6) and (7.7) imply that the vector (x_0, y_0) is a convex combination of the vectors $(x_t, \dot{x}_t)$, $t \in [0, \infty)$. Since $_0 x$ is a feasible growth path in (T, A), we must have $(x_t, \dot{x}_t) \in T^0$ for almost every $t \in [0, \infty)$.[3] Strict concavity of the model (T, A) implies that $(x_0, y_0) \in T^0$ and that

$$A((x_0, y_0), W(x_0)) > \int_0^\infty \lambda(t) A((x_t, \dot{x}_t), W(x_0)) \, dt.$$

---

[3] Recall the definition of T^0 from Equation (5.14).

The strict inequality holds because of the assumption that $_0x$ is not constant on any interval of the form $[0, \epsilon)$ with $\epsilon > 0$ and because $\lambda(t)$ is strictly positive for all sufficiently small $t > 0$. The latter result follows immediately from conditions (7.1) and (7.2). Theorem 6.1 implies that there exists a function $G : T^0 \mapsto \mathbf{R} \cup \{-\infty, +\infty\}$ such that the above inequality can be rewritten as follows:

$$
\begin{aligned}
G(x_0, y_0) &- W'(x_0; y_0) \\
&> \int_0^\infty \lambda(t) \Big[G(x_t, \dot{x}_t) - W'(x_t; \dot{x}_t) \\
&\qquad + A((x_t, \dot{x}_t), W(x_0)) - A((x_t, \dot{x}_t), W(x_t)) \Big] \, dt.
\end{aligned}
$$

Because $_0x$ is an almost interior path with $x_0 \in \text{int } X^0$, Theorem 6.1 also shows that $G(x_t, \dot{x}_t) = 0$ for almost every $t \in [0, \infty)$ and that $G(x_0, y_0) \leq 0$. Therefore, the above inequality implies that

$$
\begin{aligned}
-W'(x_0; y_0) &\qquad\qquad\qquad\qquad\qquad\qquad\qquad\qquad (7.9) \\
&> \int_0^\infty \lambda(t) \Big[-W'(x_t; \dot{x}_t) + A((x_t, \dot{x}_t), W(x_0)) - A((x_t, \dot{x}_t), W(x_t)) \Big] \, dt.
\end{aligned}
$$

From the definition of the maximal discount rate $\rho = \bar{\rho}(D)$ (see Definition 5.7) and from assumption (7.8) we see that for all $t \in [0, \infty)$ the following condition is true:

$$
A((x_t, \dot{x}_t), W(x_0)) - A((x_t, \dot{x}_t), W(x_t)) \geq \rho \Big[W(x_t) - W(x_0) \Big].
$$

Upon substituting this into Formula (7.9) we see that the following inequality must hold:

$$
W'(x_0; y_0) < \int_0^\infty \lambda(t) \Big[W'(x_t; \dot{x}_t) - \rho W(x_t) + \rho W(x_0) \Big] \, dt \qquad (7.10)
$$

From condition ii) of Theorem 6.1 it follows that the function $t \mapsto W(x_t)$ is absolutely continuous and from Lemma 6.2 we can conclude that its derivative satisfies

$$
\frac{d}{dt} W(x_t) \geq W'(x_t; \dot{x}_t) \qquad \text{a.e..} \qquad (7.11)
$$

Let us denote by t_k, $k \in \mathbf{Z}^+$, the possible points of discontinuity of the function μ and define $t_0 = 0$. From the definition of the function λ we can conclude that the identity

$$
\dot{\lambda}(t) = e^{-\rho t} \dot{\mu}(t) - \rho \lambda(t)
$$

holds for almost all $t \in [0, \infty)$. Using this fact as well as (7.5) and (7.11) we obtain from (7.10) by partial integration

$$
\begin{aligned}
W'(x_0; y_0) \\
< \rho W(x_0) &\int_0^\infty \lambda(t)\, dt \\
+ \sum_{k=1}^\infty &\int_{t_{k-1}}^{t_k} \left[\lambda(t)\frac{d}{dt}W(x_t) + \dot\lambda(t)W(x_t) - e^{-\rho t}\mu(t)W(x_t) \right] dt \\
= \rho W(x_0) & \\
+ \sum_{k=1}^\infty &\left[\lambda(t_k^-)W(x_{t_k}) - \lambda(t_{k-1}^+)W(x_{t_{k-1}}) - \int_{t_{k-1}}^{t_k} e^{-\rho t}\dot\mu(t)W(x_t)\, dt \right]
\end{aligned}
$$

where we have denoted by $\lambda(t_k^-)$ and $\lambda(t_k^+)$ the right hand side limit and the left hand side limit, respectively, of the function λ at the point t_k.[4] By substituting the definition of $\lambda(t)$ and by rearranging terms in the infinite sum we obtain[5]

$$
\begin{aligned}
W'(x_0; y_0) < (\rho - \mu(0))W(x_0) \\
+ \sum_{k=1}^\infty \left\{ e^{-\rho t_k} \left[\mu(t_k^-) - \mu(t_k^+) \right] W(x_{t_k}) - \int_{t_{k-1}}^{t_k} e^{-\rho t}\dot\mu(t)W(x_t)\, dt \right\}.
\end{aligned}
$$

It follows from this inequality and from (6.1) that there exists a subgradient $x^* \in \partial W(x_0)$ such that

$$
\begin{aligned}
x^* \cdot y_0 < (\rho - \mu(0))W(x_0) \\
+ \sum_{k=1}^\infty \left\{ e^{-\rho t_k} \left[\mu(t_k^-) - \mu(t_k^+) \right] W(x_{t_k}) - \int_{t_{k-1}}^{t_k} e^{-\rho t}\dot\mu(t)W(x_t)\, dt \right\}.
\end{aligned} \tag{7.12}
$$

Now consider the definition of y_0, i.e., Equation (7.7). Using partial integration we see that

$$
\begin{aligned}
y_0 &= \sum_{k=1}^\infty \int_{t_{k-1}}^{t_k} e^{-\rho t}\mu(t)\dot x_t\, dt \\
&= \sum_{k=1}^\infty \left\{ e^{-\rho t_k}\mu(t_k^-)x_{t_k} - e^{-\rho t_{k-1}}\mu(t_{k-1}^+)x_{t_{k-1}} - \int_{t_{k-1}}^{t_k} e^{-\rho t}\dot\mu(t)x_t\, dt \right\} \\
&\quad + \rho \int_0^\infty e^{-\rho t}\mu(t)x_t\, dt.
\end{aligned}
$$

---

[4] Both of these limits exist according to the definition of a piecewise absolutely continuous function (see Footnote 2 on page 139).

[5] The infinite sum converges absolutely so that rearranging the terms is feasible. Absolute convergence follows from the same argument as in the proof of Theorem 4.1.

The last term on the right hand side is equal to ρx_0 by virtue of condition (7.3). Using this fact and rearranging the terms in the sum we obtain

$$y_0 = (\rho - \mu(0))x_0 + \sum_{k=1}^{\infty} \left\{ e^{-\rho t_k} \left[\mu(t_k^-) - \mu(t_k^+) \right] x_{t_k} - \int_{t_{k-1}}^{t_k} e^{-\rho t} \dot{\mu}(t) x_t \, dt \right\}.$$

Finally, substitution of this expression into (7.12) yields

$$-(\rho - \mu(0))\left[W(x_0) - x^* \cdot x_0 \right]$$

$$< \sum_{k=1}^{\infty} \left\{ e^{-\rho t_k} \left[\mu(t_k^-) - \mu(t_k^+) \right] \left[W(x_{t_k}) - x^* \cdot x_{t_k} \right] \right.$$

$$\left. - \int_{t_{k-1}}^{t_k} e^{-\rho t} \dot{\mu}(t) \left[W(x_t) - x^* \cdot x_t \right] dt \right\}. \tag{7.13}$$

On the other hand, it follows from (7.1) that $\mu(t_k^-) - \mu(t_k^+) \geq 0$ for all $k \in \mathbf{Z}^+$ and that $\dot{\mu}(t) \leq 0$ for almost every $t \in [0, \infty)$. Moreover, we obtain from condition (7.2) that

$$-(\rho - \mu(0)) = \mu(0) - \rho \sum_{k=1}^{\infty} \int_{t_{k-1}}^{t_k} e^{-\rho t} \mu(t) \, dt$$

$$= \mu(0) + \sum_{k=1}^{\infty} \left\{ \left[e^{-\rho t_k} \mu(t_k^-) - e^{-\rho t_{k-1}} \mu(t_{k-1}^+) \right] - \int_{t_{k-1}}^{t_k} e^{-\rho t} \dot{\mu}(t) \, dt \right\}$$

$$= \sum_{k=1}^{\infty} \left\{ e^{-\rho t_k} \left[\mu(t_k^-) - \mu(t_k^+) \right] - \int_{t_{k-1}}^{t_k} e^{-\rho t} \dot{\mu}(t) \, dt \right\}. \tag{7.14}$$

Concavity of W shows that for all $x^* \in \partial W(x_0)$ and for all $t \in [0, \infty)$ the following is true: $W(x_t) - x^* \cdot x_t \leq W(x_0) - x^* \cdot x_0$. From this property and from (7.14) we obtain

$$-(\rho - \mu(0))\left[W(x_0) - x^* \cdot x_0 \right]$$

$$= \sum_{k=1}^{\infty} \left\{ e^{-\rho t_k} \left[\mu(t_k^-) - \mu(t_k^+) \right] \left[W(x_0) - x^* \cdot x_0 \right] \right.$$

$$\left. - \int_{t_{k-1}}^{t_k} e^{-\rho t} \dot{\mu}(t) \left[W(x_0) - x^* \cdot x_0 \right] dt \right\}$$

$$\geq \sum_{k=1}^{\infty} \left\{ e^{-\rho t_k} \left[\mu(t_k^-) - \mu(t_k^+) \right] \left[W(x_{t_k}) - x^* \cdot x_{t_k} \right] \right.$$

$$\left. - \int_{t_{k-1}}^{t_k} e^{-\rho t} \dot{\mu}(t) \left[W(x_t) - x^* \cdot x_t \right] dt \right\}.$$

This is an obvious contradiction to (7.13) which completes the proof. □

As in the discrete time case there is a simple verbal formulation of this Minimum Impatience Theorem, namely, that the initial capital stock of any non-stationary almost interior optimal growth path in a strictly concave optimal growth model cannot be represented as the weighted average of all future capital stocks along this path, when the weights are required to converge to zero at a rate which exceeds the maximal discount rate along the path.

The requirement that $_0x$ is an almost interior optimal path seems at first glance to be a rather restrictive one. However, it does not matter at all if we happen to know in advance that $_0x$ can only be an (almost) interior path. This will usually be the case if we apply Theorem 7.1 to a trajectory which is determined by a given policy function. To explain this issue and to illustrate our result we provide the following example.

Consider the system of two non-linear differential equations

$$\dot{x}_t = x_t + y_t - x_t \left[x_t^2 + y_t^2\right] \tag{7.15}$$

$$\dot{y}_t = -x_t + y_t - y_t \left[x_t^2 + y_t^2\right]. \tag{7.16}$$

By transforming this system into polar coordinates one can easily show that there exists a unique stable limit cycle $x_t = \cos t$, $y_t = -\sin t$. More generally, the solution starting in the point (x_0, y_0) is given by (see [15, p. 444])

$$x_t = \frac{\cos(t-a)}{\sqrt{1 + (b^{-1} - 1)e^{-2t}}}$$

$$y_t = \frac{-\sin(t-a)}{\sqrt{1 + (b^{-1} - 1)e^{-2t}}}$$

where $a = \arctan(y_0/x_0)$ and $b = x_0^2 + y_0^2$. For our calculations we have chosen the trajectory $(_0x, _0y)$ starting in $x_0 = [3e^{4\pi} + 1]^{-1/2}$ and $y_0 = 0$. This trajectory passes through the point $(0.5, 0)$ at time $t = 2\pi$. Specifying the function $\mu : [0, \infty) \mapsto \mathbf{R}$ by

$$\mu(t) = \begin{cases} 1.70102 & \text{for } 0.0 \leq t < 0.5 \\ 0.44686 & \text{for } 0.5 \leq t < 4.0 \\ 0.42669 & \text{for } 4.0 \leq t < 7.0 \\ 0 & \text{otherwise} \end{cases}$$

(all figures rounded to 5 decimal places) one can check that conditions (7.1) and (7.2) are satisfied with $\rho = 0.9$ and that the equations

$$\int_0^\infty e^{-0.9t} \mu(t) x_t \, dt = x_0$$

$$\int_0^\infty e^{-0.9t}\mu(t)y_t\,dt = y_0$$

hold. Therefore, we have $(x_0, y_0) \in \hat{H}((_0x, _0y), 0.9)$ and we conclude that the given path cannot be an almost interior optimal growth path of a strictly concave optimal growth problem with a maximal discount rate along this path which is smaller than 90%.[6] We shall now show that we can omit the qualification "almost interior" from this statement. If the path $(_0x, _0y)$ is optimal in an optimal growth model (T, A) with state space $X \subseteq \mathbf{R}^2$, then it must hold that the open unit disc,

$$S = \left\{(x, y) \mid \sqrt{x^2 + y^2} < 1\right\},$$

is a subset of the set X^0 of viable initial states. This is just a simple consequence of the convexity of X^0. On the other hand, since S is an open set, this property already shows that the given path is an interior path.

The situation described in the above example is typical for the case where the path $_0x$ under consideration is determined as a solution to a system of differential equations. In that case we know that the set X^0 of viable initial states has to be convex and invariant with respect to the flow determined by the differential equations. In particular, the set X^0 must contain the smallest convex and invariant set which contains the given path $_0x$. In many cases, this information can be used to prove that $_0x$ is an interior path. In this regard we would like to point out another observation which shows that the interiority condition in Theorem 7.1 is not very restrictive. In fact, little scrutiny of the proof of the theorem reveals that we need $x_t \in \text{int } X^0$ only for those $t \in [0, \infty)$ for which $\mu(t) > 0$. In the above example, it is therefore enough to know that $x_t \in \text{int } X^0$ for almost all $t \in [0, 7]$.

The example is typical also in the sense that usually one does not only know a single optimal growth path but rather a policy function h which generates the given path as one of its trajectories. Let us therefore reformulate Theorem 7.1 in terms of a policy function h. To be more specific, let us assume that $S \subseteq X$ is a subset of X, and that $h : S \mapsto \mathbf{R}^n$ is a continuous function such that the following is true: for every $x \in S$ there exists a unique solution $_0x$ of the differential equation

$$\dot{x}_t = h(x_t) \tag{7.17}$$

subject to the initial condition $x_0 = x$ and this solution satisfies the constraint $x_t \in S$ for all $t \in [0, \infty)$. It is clear from this specification and from Definition 6.2 that h is a candidate for an optimal policy function in any optimal growth model with state space $X \supseteq S$. For example consider the function h defined

---

[6] In Section 6.3 we shall derive a better bound on the maximal discount rate for this example, namely, $\bar{\rho}(\mathbf{F}) \geq 1$.

by the right hand sides of (7.15) and (7.16). The function satisfies the above requirements if S is any (open or closed) disk with center $(0,0)$ and radius greater than or equal to one.

Analogously to the discrete time case we can associate to the function $h : S \mapsto \mathbf{R}^n$ the average correspondence $H : S \times [0, \infty) \mapsto 2^S$ in the following way.

Definition 7.1 Let $h : S \mapsto \mathbf{R}^n$ be as described before and let $x \in S$ and $\rho \geq 0$ be given. Denote by $_0x$ the unique solution of (7.17) which satisfies $x_0 = x$ and define $H(x, \rho) = \tilde{H}(_0x, \rho)$. The set-valued mapping $H : S \times [0, \infty) \mapsto 2^S$ is called the *average correspondence*.

For the reformulation of Theorem 7.1 in terms of a policy function it will be convenient to define a non-trivial fixpoint of the average correspondence as follows.

Definition 7.2 Let $H : S \times [0, \infty) \mapsto 2^S$ be the average correspondence associated with a given mapping $h : S \mapsto \mathbf{R}^n$. We say that the vector $x \in S$ is a *non-trivial fixpoint* of H at ρ if the conditions $h(x) \neq 0$ and $x \in H(x, \rho)$ are satisfied.

In the following theorem we assume that the set S on which the function h is defined is an open subset of the state space $X \subseteq \mathbf{R}^n$. By this assumption we rule out non-interior optimal growth paths which helps to simplify the exposition.

Theorem 7.2 *Let $X \subseteq \mathbf{R}^n$ be a non-empty, closed, and convex set, $S \subseteq X$ an open set, and $h : S \mapsto \mathbf{R}^n$ a given continuous function satisfying the conditions discussed above. Assume that there exists $x \in S$ and $\rho \in [0, \infty)$ such that x is a non-trivial fixpoint of the average correspondence H at ρ. Define D as the set consisting of the single path $_0x$, where $_0x$ denotes the unique trajectory of (7.17) emanating from $x_0 = x$. There does not exist any strictly concave optimal growth model (T, A) with state space X such that h is an optimal policy function for this model and such that the maximal discount rate along $_0x$ satisfies $\bar{\rho}(D) \leq \rho$.*

PROOF. The assumptions imply that $_0x$ can only be an interior optimal growth path for (T, A) if it is optimal at all. Therefore, the result follows immediately from Definitions 7.1 and 7.2 and from Theorem 7.1. □

Assume that $_0x$ is an optimal growth path in some strictly concave continuous time optimal growth model (T, A). We are now going to show briefly how one can use the time-τ transformation discussed at the end of Section 6.2 to derive a lower bound on the globally maximal discount rate of the aggregator function A by applying a Minimum Impatience Theorem for discrete time problems. To this end we pick an arbitrary period length $\tau > 0$ and define the discrete time growth path $_0 x' = (x_0, x_\tau, x_{2\tau}, \ldots)$. From Lemma 6.9 we know that there is a strictly concave discrete time optimal growth model (T', A') for which $_0 x'$ is an optimal growth path. Now assume that any Minimum Impatience Theorem from Part 1 is applicable to this new problem and that it yields a bound $\underline{\delta}(\mathbf{F'}) < \delta'$. Again referring to Lemma 6.9 it follows that the globally maximal discount rate of the original model must satisfy $\bar{\rho}(\mathbf{F}) > -(1/\tau)\ln(\delta')$.

As a first illustration of this reasoning consider our previous example (7.15), (7.16). If we choose the initial state (x_0, y_0) with $x_0 \in (0,1)$ and $y_0 = 0$ and the period length $\tau = \pi$, then we obtain $y_{k\tau} = 0$ for all $k \in \mathbf{Z}_0^+$ and

$$x_{k\tau} = \frac{(-1)^k}{\sqrt{1 + (x_0^{-2} - 1)e^{-2k\pi}}}.$$

In order to apply Theorem 4.1 we need numbers $\delta \in (0,1)$ and μ_1, μ_2, μ_3, $\ldots$ such that conditions (4.1) and (4.2) are satisfied and such that

$$x_0 = \sum_{k=1}^{\infty} \delta^k \mu_k x_{k\tau}. \tag{7.18}$$

We specify $\mu_k = 0$ for all $k \geq 3$ and $\mu_1 = \mu_2 = (\delta + \delta^2)^{-1}$. This implies immediately that (4.1) and (4.2) are satisfied. Substituting the values for μ_k, $k \in \mathbf{Z}^+$, into Equation (7.18) and rearranging terms we obtain

$$\delta = \left[\frac{1}{\sqrt{x_0^2 + (1 - x_0^2)e^{-2\pi}}} + 1 \right] \Big/ \left[\frac{1}{\sqrt{x_0^2 + (1 - x_0^2)e^{-4\pi}}} - 1 \right].$$

In the limit as x_0 approaches zero we obtain

$$\delta = \frac{1 + e^\pi}{e^{2\pi} - 1}$$

which shows that the minimum rate of impatience required for the optimality of the system (7.15) - (7.16) is bounded below by

$$-\frac{1}{\pi} \ln \frac{1 + e^\pi}{e^{2\pi} - 1} \approx 0.986.$$

This bound is better than the one we have derived earlier from Theorem 7.1 although it can be computed almost by hand without the need to do numerical integrations.

A second way to use the time-τ transformation is to combine it with Theorem 4.4. This approach leads to the following result.

Theorem 7.3 *Let $X \subseteq \mathbf{R}^n$ be a non-empty, closed, and convex set and let $_0\bar{x} \in \mathcal{AC}(X)$ and $_0x \in \mathcal{AC}(X)$ be given functions. Assume that $\bar{x}_0 \neq x_0$ and that there exist numbers $\tau > 0$ and $\rho > 0$ such that the following conditions are satisfied:*

i) $\bar{x}_\tau$ is in the relative interior of X,

ii) $\bar{x}_\tau = \bar{x}_0$, and

iii) $x_0 - \bar{x}_0 = e^{-\rho\tau}(x_\tau - \bar{x}_\tau)$.

If both trajectories $_0\bar{x}$ and $_0x$ are optimal growth paths in the same strictly concave optimal growth model (T, A), then it holds that the globally maximal discount rate of A satisfies $\bar{\rho}(\mathbf{F}) > \rho$.

PROOF. Consider the discrete time optimal growth model (T', A') obtained from (T, A) by the time-τ transformation. From Lemma 6.9 it follows that the sequences $(\bar{x}_0, \bar{x}_\tau, \bar{x}_{2\tau}, \ldots)$ and $(x_0, x_\tau, x_{2\tau}, \ldots)$ are optimal growth paths for (T', A') and that $\underline{\delta}(\mathbf{F}') \geq e^{-\tau\bar{\rho}(\mathbf{F})}$. It is now easy to verify that all the assumptions of Theorem 4.4 are satisfied for the model (T', A') so that we conclude that $e^{-\tau\rho} > \underline{\delta}(\mathbf{F}') \geq e^{-\tau\bar{\rho}(\mathbf{F})}$. Obviously, this implies that $\bar{\rho}(\mathbf{F}) > \rho$ and the proof is complete. □

Contrary to Theorem 7.1 the above Minimum Impatience Theorem can be applied to optimal growth paths which are not necessarily non-monotonic. In particular, Theorem 7.3 is also useful for one-dimensional optimal growth models. This is illustrated by the following example.

Consider the state space $X = [-\pi/2, \pi/2]$ and the dynamical system

$$\dot{x}_t = h(x_t) = B \sin x_t \cos x_t. \tag{7.19}$$

We claim that $\rho = B$ is a lower bound on the maximal discount rate $\bar{\rho}(\mathbf{F})$ of any strictly concave optimal growth model which has (7.19) as its optimal policy function. The trajectories of (7.19) are given by

$$x_t = \begin{cases} -\pi/2 & \text{for } x_0 = -\pi/2 \\ \arctan(e^{Bt} \tan x_0) & \text{for } x_0 \in (-\pi/2, \pi/2) \\ \pi/2 & \text{for } x_0 = \pi/2 \end{cases}$$

which shows that there are three stationary point $-\pi/2$, 0, and $\pi/2$. Let us chose $\bar{x}_0 = 0$ so that conditions i) and ii) of Theorem 7.3 are both satisfied for all $\tau > 0$. For a given $x_0 \in (0, \pi/2)$ condition iii) is also satisfied provided that

$$e^{\rho\tau} = \frac{x_\tau}{x_0} = \frac{\arctan(e^{B\tau} \tan x_0)}{x_0}.$$

It is straightforward to check that the limit of the right hand side of this equation as x_0 approaches zero is given by $e^{B\tau}$. This proves the claim that no strictly concave optimal growth model with a maximal discount rate less than or equal to B can have the mapping (7.19) as its optimal policy function.

We conclude this section by pointing out that most of the remarks made with regard to Theorems 4.1 - 4.4 carry over with obvious modifications to the continuous time Minimum Impatience Theorems. In particular, we can determine the best bound achievable by Theorem 7.2 as the solution of the following non-linear and infinite dimensional optimization problem:

$$\text{Maximize } \rho$$
$$\text{subject to } x \in S, \ h(x) \neq 0, \text{ and } x \in H(x, \rho)$$

The control variables in this problem are x, ρ, and the functions μ used in conditions (7.1) - (7.3). Solving this problem, however, is even harder than in the discrete time case. The most promising approach seems to be an approximation based on a finite dimensional parametric subfamily of all possible functions μ. Such a family could be defined in terms of piecewise constant functions or piecewise linear functions. Finally, it is clear that knowledge of the complete trajectory $_0x$ is not necessary for the application of Theorem 7.1. If we have access only to a finite initial segment, $x : [0, s] \mapsto X$, then all we have to do is to set $\mu(t) = 0$ for all $t > s$. The reader will note that we used this approach in the example discussed earlier (Equations (7.15) and (7.16)) although the complete trajectory $_0x$ was known.

Analogously, one can find the best bound that can be obtained from Theorem 7.3 by solving the following non-linear program:

$$\text{Minimize } e^{-\rho\tau}$$
$$\text{subject to } x \in X, \ \bar{x} \in X, \ \tau \in [0, \infty),$$
$$h^{(\tau)}(\bar{x}) = \bar{x},$$
$$e^{-\rho\tau}(h^{(\tau)}(x) - \bar{x}) = x - \bar{x} \neq 0.$$

7.2   The Average Correspondence

It is evident that the average correspondence plays the same pre-dominant role in Theorem 7.2 as it did in the discrete time Minimum Impatience Theorem 4.2. Let us therefore briefly discuss some basic properties of H. As a first step towards a better understanding of H we develop a representation of $\tilde{H}(_0x, \rho)$ which is more convenient than the one presented in conditions (7.1) - (7.3). For an arbitrary path $_0x$ and for every $s \in (0, \infty)$ we define

$$g_s(_0x, \rho) = \begin{cases} \dfrac{\rho}{1 - e^{-\rho s}} \displaystyle\int_0^s e^{-\rho t} x_t \, dt \,, & \text{for } \rho > 0 \\[3mm] \dfrac{1}{s} \displaystyle\int_0^s x_t \, dt \,, & \text{for } \rho = 0 \end{cases} \qquad (7.20)$$

and

$$\tilde{H}_s(_0x, \rho) = \text{conv}\Big\{ g_t(_0x, \rho) \,\Big|\, 0 < t \le s \Big\}$$

where conv C denotes the convex hull of a set C. The proof of the following result is similar to the one of Lemma 4.8 and will be omitted.

Lemma 7.2 *Let $_0x$ be an arbitrary path and let ρ, s, and s' be real numbers with $\rho \ge 0$, and $0 < s < s'$. Then it holds that $\tilde{H}_s(_0x, \rho) \subseteq \tilde{H}_{s'}(_0x, \rho) \subseteq \tilde{H}(_0x, \rho)$.*

As in the preceding section we assume that $h : S \mapsto \mathbf{R}^n$ is a continuous function such that the differential equation

$$\dot{x}_t = h(x_t) \qquad (7.21)$$

has a unique solution $_0x$ for every initial state $x_0 \in S$ and such that this solution satisfies $x_t \in S$ for all $t \in [0, \infty)$. We are going to show that we can approximate the average correspondence $H(x, \rho)$ by the increasing sequence of sets $\tilde{H}_s(_0x, \rho)$ on a suitable subset of its domain. To specify this subset we define the number β_h as the infimum of all real numbers $\beta > 0$ for which it holds that

$$\inf_{\alpha > 0} \sup \Big\{ |h(x)|/(\alpha + |x|) \,\Big|\, x \in S \Big\} \le \beta.$$

It is obvious from this definition that h cannot be the optimal policy function in an optimal growth model (T, A) with a β-bounded technology provided that $\beta < \beta_h$. An application of the Minimum Impatience Theorems to exclude the optimality of the policy function h is therefore only necessary if $\beta \ge \beta_h$ which, in view of condition iii) of Definition 5.3, implies that $\underline{\rho} > \beta_h$. The relevant domain on which we have to consider the average correspondence is therefore given by $S \times (\beta_h, \infty)$. On this domain we have the following approximation result.

Theorem 7.4 *For every $x \in S$ and for every $\rho \in (\beta_h, \infty)$ it holds that*

$$\bigcup_{s > 0} \tilde{H}_s(_0x, \rho) \subseteq H(x, \rho) \subseteq \overline{\bigcup_{s > 0} \tilde{H}_s(_0x, \rho)} \qquad (7.22)$$

where the bar denotes the closure operation and where $_0x$ denotes the unique trajectory of (7.21) emanating from the initial state $x_0 = x$.

PROOF. The left hand inclusion follows from Lemma 7.2. The proof of the right hand inclusion is very similar to the proof of (4.32) in Theorem 4.5 and uses the fact that the integrals in (7.2) and (7.3) converge absolutely. We omit the details. □

It should be noted that Theorem 7.4 is much weaker than the corresponding result in discrete time (Theorem 4.5) since we have not shown that the average correspondence is closed. In fact, if we were able to prove that $H(x, \rho)$ is closed for some $x \in S$ with $h(x) \neq 0$, then it would already follow that x is a non-trivial fixpoint of H at ρ. This is an immediate consequence of the following lemma.

Lemma 7.3 *For every $x \in S$ and for every $\rho \geq 0$ it holds that $x \in \overline{H(x, \rho)}$.*

PROOF. From the definition of $g_s(_0x, \rho)$ it follows that $\lim_{s \searrow 0} g_s(_0x, \rho) = x_0$ holds for all $\rho \in [0, \infty)$. Because this obviously implies that

$$x_0 \in \overline{\bigcup_{s>0} \tilde{H}_s(_0x, \rho)},$$

the result follows from Theorem 7.4. □

We conclude the section by summarizing some of the properties of the average correspondence.

Lemma 7.4 *Let S be a given set and $h : S \mapsto \mathbf{R}^n$ a continuous function such that solutions of (7.21) are uniquely determined by their initial conditions and satisfy $x_t \in S$ for all $t \in [0, \infty)$. For the associated average correspondence $H : S \times [0, \infty) \mapsto 2^S$ the following properties are true:*

i) $H(x, \rho) \subseteq H(x, \rho')$ for all $0 \leq \rho' \leq \rho$ and all $x \in S$.

ii) $\lim_{\rho \to \infty} \overline{H(x, \rho)} = \{x\}$ for all $x \in S$.

iii) $\overline{H(x, \rho)}$ is a singleton if and only if $h(x) = 0$. In that case the following identities hold true: $\overline{H(x, \rho)} = H(x, \rho) = \{x\}$.

PROOF. Statement i) follows from Lemma 7.1. Statement ii) follows from the observation that $\lim_{\rho \to \infty} \tilde{H}_s(_0x, \rho) = \{x_0\}$ and from Theorem 7.4. Statement iii) follows immediately from Definition 7.1. □

7.3 Local Conditions

In the final section of this chapter we show how one can use local properties of an optimal policy function h at a stationary or periodic point to derive lower bounds on the maximal discount rate along certain paths generated by h. As in the discrete time case we can derive such local Minimum Impatience Theorems from both types of global results, i.e., from Theorem 7.1 and Theorem 7.3. We can capture the main idea underlying the first approach by the notion of an expanding stationary state. In order to do this we first have to define what we mean by a ρ-expanding matrix.[7]

Definition 7.3 Let J be a real $n \times n$ matrix and let $\rho \in [0, \infty)$ be a real number. The matrix J is said to be ρ-expanding (relative to $\mathbf{R}^n$) if there exists a vector $u \in \mathbf{R}^n$ such that the following is true: for all real numbers $\gamma > 0$ there exists a number $s(\gamma) \in [0, \infty)$ such that

$$B(0, \gamma) \subseteq \text{conv} \left\{ e^{-\rho t} e^{Jt} u \,\middle|\, 0 \leq t \leq s(\gamma) \right\}.$$

Note that e^{Jt} is the $n \times n$ matrix defined by the power series

$$e^{Jt} = \sum_{k=0}^{\infty} J^k \frac{t^k}{k!}.$$

We shall need the following result which can be proved in the same way as Lemma 4.10.

Lemma 7.5 Let J be a real $n \times n$ matrix and $\rho \in [0, \infty)$ a real number. If the matrix J is ρ-expanding and if λ is an eigenvalue of J, then it holds that the real part of λ is strictly greater than ρ. In particular, the matrices J and $J - \rho I$ are non-singular.

As in the discrete time case we call a stationary point $\bar{x}$ of the system

$$\dot{x}_t = h(x_t) \tag{7.23}$$

ρ-expanding if the Jacobian matrix $J = h'(\bar{x})$ is ρ-expanding. Again we assume that the domain S of the function h is an open set to rule out non-interior growth paths.

---

[7]In a similar way as in Definition 4.3 we could define the term "ρ-expanding relative to a given set C". However, there is no need to do so because we shall only consider interior stationary states. For such points the cone of interior directions coincides with $\mathbf{R}^n$.

Definition 7.4 Let $S \subseteq \mathbf{R}^n$ be an open set and let $h : S \mapsto \mathbf{R}^n$ be a continuous function such that solutions of (7.23) are uniquely determined by their initial conditions and satisfy $x_t \in S$ for all $t \in [0, \infty)$. Moreover, let $\rho \in [0, \infty)$ be an arbitrary number and $\bar{x} \in S$ a vector such that $h(\bar{x}) = 0$. The point $\bar{x}$ is called a *ρ-expanding* fixpoint of h, if the function h is continuously differentiable in a neighborhood of $\bar{x}$ and if the Jacobian matrix $J = h'(\bar{x})$ of h at $\bar{x}$ is ρ-expanding in the sense of Definition 7.3.

In the following theorem we derive a connection between the existence of a ρ-expanding fixpoint of h and the minimum rate of impatience required for the optimality of h as an optimal policy function in a strictly concave optimal growth model.

Theorem 7.5 *Let $S \subseteq \mathbf{R}^n$ be an open set and let $h : S \mapsto \mathbf{R}^n$ be a continuous function such that solutions of (7.23) are uniquely determined by their initial conditions and satisfy $x_t \in S$ for all $t \in [0, \infty)$. Moreover, let $\rho \in [0, \infty)$ be an arbitrary number and $\bar{x} \in S$ a ρ-expanding fixpoint of h. Then it follows that there exists a non-trivial fixpoint of the average correspondence H at ρ. Consequently, there does not exist a strictly concave optimal growth model (T, A) such that h is an optimal policy function for this model and such that the globally maximal discount rate satisfies $\bar{\rho}(\mathbf{F}) \le \rho$.*

PROOF. The proof is similar to the one of Theorem 7.5 and will only be outlined. Let $u \in \mathbf{R}^n$ and $\epsilon > 0$ be given and denote by $_0 x^{(u,\epsilon)}$ the unique solution of (7.23) subject to the initial condition $x_0 = \bar{x} + \epsilon u$. Because of the assumed differentiability of h we have

$$x_t^{(u,\epsilon)} = \bar{x} + \epsilon e^{Jt} u + o(\epsilon).$$

Here, $\lim_{\epsilon \to 0} o(\epsilon)/\epsilon = 0$ uniformly in t as long as t is constrained to be in a compact interval. This implies

$$g_s\big(_0 x^{(u,\epsilon)}, \rho\big) = \bar{x} + \frac{\epsilon \rho}{1 - e^{-\rho s}} \int_0^s e^{-\rho t} e^{Jt} u \, dt + o(\epsilon).$$

Integration yields the following equation which is the exact counterpart to (4.36):

$$g_s\big(_0 x^{(u,\epsilon)}, \rho\big) = \bar{x} - \frac{\epsilon \rho}{1 - e^{-\rho s}}(J - \rho I)^{-1} u + \frac{\epsilon \rho}{1 - e^{-\rho s}}(J - \rho I)^{-1} e^{-\rho s} e^{Js} u + o(\epsilon)$$

From this point onwards the proof proceeds along the same lines as in Theorem 4.6. One can show that there exists $\bar{u} \in \mathbf{R}^n$ such that $\bar{x} + \epsilon \bar{u}$ is a fixpoint of the average correspondence H at ρ for all sufficiently small $\epsilon > 0$. If all of these fixpoints were trivial ones, then it would follow that $\lambda = 0$ is among the eigenvalues of J which is a contradiction to Lemma 7.5. □

We have already mentioned in the introduction to the present chapter that the continuous time Minimum Impatience Theorem 7.1 is of no use in one-dimensional growth models because one-dimensional differential equations can only have monotonic solutions. This becomes even more evident from Theorem 7.5 which was derived from Theorem 7.1. As a matter of fact, it is easy to see that a function $h : S \mapsto \mathbf{R}$ with $S \subseteq \mathbf{R}$ cannot have a ρ-expanding fixpoint for any $\rho \in [0, \infty)$. More generally, a little thought reveals that a fixpoint cannot be ρ-expanding if one of the eigenvalues of the Jacobian is real. This implies that Theorem 7.5 is not applicable to a policy function h defined on a state space with an odd dimension.[8]

For the case $n = 2$ it turns out that a fixpoint $\bar{x}$ is ρ-expanding if and only if the Jacobian matrix J has a pair of complex conjugate eigenvalues $\lambda = \sigma \pm i\omega$ with $\sigma > \rho$ and $\omega \neq 0$. As an example let us consider the two-dimensional system (7.15) - (7.16) discussed in Section 7.1. The Jacobian matrix J of this system evaluated at the fixpoint $(0,0)$ is given by

$$ J = \begin{pmatrix} 1 & 1 \\ -1 & 1 \end{pmatrix} $$

and has the eigenvalues $\lambda = 1 \pm i$. This shows that $(0,0)$ is ρ-expanding for all $\rho < 1$ and it follows that the system (7.15) - (7.16) cannot represent the optimal policy function in any strictly concave optimal growth model with a discount rate smaller than 100%. This bound is better than those presented in Section 6.1 although it uses only the local properties of the underlying function h at the stationary point $(0,0)$.

In the continuous time framework we were not able to use the time-τ transformation to generalize Theorem 7.5 for τ-periodic trajectories. To explain the difficulties let us assume that the optimal policy function h of a strictly concave optimal growth model (T, A) admits a closed orbit of period τ. This means that there is a trajectory $_0\bar{x}$ of (7.23) such that $\bar{x}_t = \bar{x}_{t+\tau}$ for all $t \in [0, \infty)$. If we apply the time-τ transformation to the model (T, A), we obtain a strictly concave discrete time optimal growth model (T', A') which has the time-τ map $h^{(\tau)}$ as its optimal policy function. Every point on the closed orbit of h is a fixpoint of the function $h^{(\tau)}$ but none of them is δ-expanding in the sense of Definition 4.4 for some $\delta < 1$. This follows simply from the fact that the eigenvalues of the Jacobian of $h^{(\tau)}$ are exactly the Floquet multipliers (characteristic roots) of the

---

[8]This does not imply, however, that the global Minimum Impatience Theorems 7.1 and 7.2 are not conclusive on odd-dimensional state spaces.

τ-periodic variational equation

$$\dot{y}(t) = h'(\bar{x}_t)y(t). \tag{7.24}$$

Since one of these multipliers is $\lambda = 1$ (see, e.g., [4, p. 344] or [24, p. 352]), it can be seen from Lemma 4.10 that the fixpoint is not δ-expanding for any $\delta < 1$.

To be able to handle periodic orbits and policy functions defined on odd-dimensional state spaces we must therefore derive a local Minimum Impatience Theorem from Theorem 7.3 or directly from Theorem 4.8. This leads to the following result.

Theorem 7.6 *Let $X \subseteq \mathbf{R}^n$ be a non-empty, closed, and convex set and S a closed and convex subset of X. Moreover, let $h : X \mapsto \mathbf{R}^n$ be a continuously differentiable function such that S is an invariant set for the flow defined by*

$$\dot{x}_t = h(x_t). \tag{7.25}$$

Assume that $\bar{x}_0$ is in the relative interior of S and that there exist a number $\tau > 0$ and a trajectory $_0\bar{x}$ of system (7.25) such that $_0\bar{x}$ is either stationary or τ-periodic. Denote by $\Psi(t)$ a fundamental matrix for the variational equation (7.24) satisfying $\Psi(0) = I$. If $\Psi(\tau)$ has the positive real eigenvalue λ with algebraic multiplicity one then it follows that h cannot be the optimal policy function of any strictly concave optimal growth model (T, A) with state space X such that the globally maximal discount rate satisfies $\bar{\rho}(\mathbf{F}) \leq \tau^{-1} \ln |\lambda|$. The result remains true if λ is a negative eigenvalue of $\Psi(\tau)$ with multiplicity one provided that $-\lambda$ is not an eigenvalue of $\Psi(\tau)$.

PROOF. As in the proof of Theorem 7.3 we consider the discrete time optimal growth model (T', A') obtained from (T, A) by the time-τ transformation. Note that we may assume that (T', A') has the state space S. It follows that the time-τ map $h^{(\tau)}$ is the optimal policy function of (T', A') and that $\bar{x}_0$ is a fixpoint of $h^{(\tau)}$. The result follows now from Theorem 4.8 and the fact that the Jacobian of the map $h^{(\tau)}$ is given by the fundamental matrix $\Psi(\tau)$. $\qquad \square$

As an example consider one of the most famous continuous time dynamical systems exhibiting chaos, i.e., the Lorenz model (see [43] or [33]). It is given by the three-dimensional system

$$\dot{x}_t = 10(y_t - x_t)$$
$$\dot{y}_t = 28x_t - y_t - x_t z_t$$
$$\dot{z}_t = x_t y_t - (8/3)z_t$$

The point $(\bar{x}_0, \bar{y}_0, \bar{z}_0) = (0, 0, 0)$ is a stationary point and the eigenvalues of the Jacobian at this point are given by $-8/3$, -22.82772, and 11.82772, respectively. It follows that $e^{11.82772\tau}$ is an eigenvalue of the fundamental matrix

$\Psi(\tau)$ corresponding to (7.24). Consequently, we have $\bar{\rho}(\mathbf{F}) > 11.82772$ for every strictly concave optimal growth model which has the Lorenz equations as its optimal policy function and for which the origin belongs to the relative interior of the state space X.

Bibliography

[1] J. P. AUBIN, *Mathematical Methods of Game and Economic Theory*, North-Holland, Amsterdam, 1979.

[2] J. P. AUBIN, A survey of viability theory, *SIAM Journal of Control and Optimization* **28** (1990), pp. 749 – 788.

[3] J. P. AUBIN AND A. CELLINA, *Differential Inclusions*, Springer Verlag, Berlin, 1984.

[4] H. AMANN, *Gewöhnliche Differentialgleichungen*, de Gruyter, Berlin, 1983.

[5] W. A. BARNETT, J. GEWEKE, AND K. SHELL (Eds.), *Economic Complexity: Chaos, Sunspots, Bubbles, and Nonlinearity*, Cambridge University Press, Cambridge, 1989.

[6] R. F. BAUM, Existence theorems for Lagrange control problems with unbounded time domain, *Journal of Optimization Theory and Applications* **19** (1976), pp. 89 - 116.

[7] R. A. BECKER, J. H. BOYD III, AND B. Y. SUNG, Recursive utility and optimal capital accumulation. I. Existence, *Journal of Economic Theory* **47** (1989), pp. 76 - 100.

[8] J. BENHABIB AND K. NISHIMURA, Competitive equilibrium cycles, *Journal of Economic Theory* **35** (1985), pp. 284 - 306.

[9] J. BENHABIB AND A. RUSTICHINI, Equilibrium cycling with small discounting, *Journal of Economic Theory* **52** (1990), pp. 423 - 432.

[10] C. BERGE, *Espaces Topologiques, Fonctions Multivoques*, Dunod, Paris, 1959.

[11] M. BOLDRIN, Persistent oscillations and chaos in dynamic economic models: notes for a survey, in *The Economy as an Evolving Complex System* (P. W. Anderson, K. J. Arrow, and D. Pines, Eds.), pp. 49 - 75, Addison-Wesley, Reading MA, 1988.

158

[12] M. BOLDRIN AND R. DENECKERE, Sources of complex dynamics in two-sector growth models, *Journal of Economic Dynamics and Control* **14** (1990), pp. 627 - 653.

[13] M. BOLDRIN AND L. MONTRUCCHIO, On the indeterminacy of capital accumulation paths, *Journal of Economic Theory* **40** (1986), pp. 26 - 39.

[14] M. BOLDRIN AND M. WOODFORD, Equilibrium models displaying endogenous fluctuations and chaos, *Journal of Monetary Economics* **25** (1990), pp. 189 - 222.

[15] W. E. BOYCE AND R. C. DiPRIMA, *Elementary Differential Equations and Boundary Value Problems*, 3rd Ed., John Wiley & Sons, New York NY, 1977.

[16] J. H. BOYD III, Recursive utility and the Ramsey problem, *Journal of Economic Theory* **50** (1990), pp. 326 - 345.

[17] W. A. BROCK AND A. G. MALLIARIS, *Differential Equations, Stability and Chaos in Dynamic Economics*, North-Holland, Amsterdam, 1989.

[18] W. A. BROCK AND J. A. SCHEINKMAN, Global asymptotic stability of optimal control systems with applications to the theory of economic growth, *Journal of Economic Theory* **12** (1976), pp. 164 - 190.

[19] D. CASS AND K. SHELL, The structure and stability of competitive dynamical systems, *Journal of Economic Theory* **12** (1976), pp. 31 - 70.

[20] L. CESARI, *Optimization – Theory and Applications*, Springer Verlag, New York NY, 1983.

[21] F.-R. CHANG, The inverse optimal problem: a dynamic programming approach, *Econometrica* **56** (1988), pp. 147 - 172.

[22] C. CHIARELLA, *The Elements of a Nonlinear Theory of Economic Dynamics*, Springer Verlag, Berlin, 1990.

[23] F. H. CLARKE, *Optimization and Nonsmooth Analysis*, John Wiley & Sons, New York NY, 1983.

[24] E. A. CODDINGTON AND N. LEVINSON, *Theory of Ordinary Differential Equations*, McGraw-Hill, New York NY, 1955.

[25] J. H. CURRY, On the Hénon transformation, *Communications in Mathematical Physics* **68** (1979), pp. 129 - 140.

[26] R.-A. DANA AND C. LE VAN, Optimal growth and Pareto optimality, *Journal of Mathematical Economics* **20** (1991), pp. 155 - 180.

[27] R. DENECKERE AND S. PELIKAN, Competitive chaos, *Journal of Economic Theory* **40** (1986), pp. 13 - 25.

[28] R. L. DEVANEY, *An Introduction to Chaotic Dynamical Systems*, Benjamin Cummings, Menlo Park CA, 1986.

[29] J.-P. ECKMANN AND D. RUELLE, Ergodic theory of chaos and strange attractors, *Reviews of Modern Physics* **57** (1985), pp. 617 - 656.

[30] L. G. EPSTEIN, The global stability of efficient intertemporal allocations, *Econometrica* **55** (1987), pp. 329 - 355.

[31] M. FRANK AND T. STENGOS, Chaotic dynamics in economic time-series, *Journal of Economic Surveys* **2** (1988), pp. 103 - 133.

[32] J.-M. GRANDMONT, *Nonlinear Economic Dynamics*, Academic Press, Orlando FL, 1987.

[33] J. GUCKENHEIMER AND P. HOLMES, *Nonlinear Oscillations, Dynamical Systems, and Bifurcations of Vector Fields*, Springer Verlag, New York NY, 1983.

[34] M. HÉNON, A two-dimensional mapping with a strange attractor, *Communications in Mathematical Physics* **50** (1976), pp. 69 - 77.

[35] T. U. HEWAGE AND D. A. NEUMANN, Functions not realizable as policy functions in an optimal growth model, Working Paper, Bowling Green State University, Bowling Green OH, 1990.

[36] W. HILDENBRAND, *Core and Equilibria of a Large Economy*, Princeton University Press, Princeton NJ, 1974.

[37] C. H. HOMMES, *Chaotic Dynamics in Economic Models: Some Simple Case-Studies*, Wolters-Noordhoff, Groningen, 1991.

[38] A. D. IOFFE AND V. M. TIHOMIROV, *Theory of Extremal Problems*, North-Holland, Amsterdam, 1979.

[39] T. C. KOOPMANS, Stationary ordinal utility and impatience, *Econometrica* **28** (1960), pp. 287 - 309.

[40] T. C. KOOPMANS, P. A. DIAMOND, AND R. E. WILLIAMSON, Stationary ordinal utility and time perspective, *Econometrica* **32** (1964), pp. 82 - 100.

[41] M. KURZ, On the inverse optimal problem, in *Mathematical Systems Theory and Economics I* (H. W. Kuhn and G. P. Szegö, Eds.), pp. 189 - 201, Springer Verlag, Berlin, 1969.

[42] T. LI AND J. A. YORKE, Period three implies chaos, *American Mathematical Monthly* **82** (1975), pp. 985 - 992.

[43] E. N. LORENZ, Deterministic nonperiodic flow, *Journal of the Atmospheric Sciences* **20** (1963), pp. 130 - 141.

[44] H. W. LORENZ, *Nonlinear Dynamical Economics and Chaotic Motion*, Springer Verlag, Berlin, 1989.

[45] R. E. LUCAS, JR., AND N. L. STOKEY, Optimal growth with many consumers, *Journal of Economic Theory* **32** (1984), pp. 139 - 171.

[46] R. M. MAY, Simple mathematical models with very complicated dynamics, *Nature* **261** (1976), pp. 459 - 467.

[47] L. W. MCKENZIE, Optimal economic growth, turnpike theorems and comparative dynamics, in *Handbook of Mathematical Economics*, Vol. 3 (K. J. Arrow and M. D. Intriligator, Eds.), pp. 1281 - 1355, North-Holland, Amsterdam, 1986.

[48] L. MONTRUCCHIO, The occurrence of erratic fluctuations in models of optimization over infinite horizon, in *Growth Cycles and Multisectoral Economics: the Goodwin Tradition* (G. Ricci and K. Velupillai, Eds.), pp. 83 - 92, Springer Verlag, Berlin, 1988.

[49] L. MONTRUCCHIO, Dynamical systems that solve continuous-time concave optimization problems: anything goes, Paper presented at the 6th IFAC Symposium on *Dynamic Modelling and Control of National Economics*, Edinburgh, 1989.

[50] L. MONTRUCCHIO, Topics on dynamics in infinite horizon concave problems, Paper presented at the 16th GMÖOR Symposium on *Operations Research*, Trier, 1991.

[51] D. NEUMANN, T. O'BRIEN, J. HOAG, AND K. KIM, Policy functions for capital accumulation paths, *Journal of Economic Theory* **46** (1988), pp. 205 - 214.

[52] K. NISHIMURA AND M. YANO, Non-linear dynamics and chaos in optimal growth: characterizations, Working Paper, Kyoto University, 1991.

[53] K. NISHIMURA AND M. YANO, Non-linear dynamics and chaos in optimal growth: an example, Working Paper, Kyoto University, 1991.

[54] T. PUU, *Nonlinear Economic Dynamics*, Springer Verlag, Berlin, 1989.

[55] A. W. ROBERTS AND D. E. VARBERG, Another proof that convex functions are locally Lipschitz, *American Mathematical Monthly* **81** (1974), pp. 1014 - 1016.

[56] R. T. ROCKAFELLAR, *Convex Analysis*, Princeton University Press, Princeton NJ, 1970.

[57] R. T. ROCKAFELLAR, Saddle points of Hamiltonian systems in convex Lagrange problems having a non-zero discount rate, *Journal of Economic Theory* **12** (1976), pp. 71 - 113.

[58] A. N. SARKOVSKII, Coexistence of cycles of a continuous map of a line into itself, *Ukrainian Mathematical Journal* **16** (1964), pp. 61 - 71.

[59] J. A. SCHEINKMAN, On optimal steady states of n-sector growth models when utility is discounted, *Journal of Economic Theory* **12** (1976), pp. 11 - 30.

[60] G. SORGER, On the optimality and stability of competitive paths in continuous time growth models, *Journal of Economic Theory* **48** (1989), pp. 526 - 547.

[61] G. SORGER, On the optimality of given feedback controls, *Journal of Optimization Theory and Applications* **65** (1990), pp. 321 - 329.

[62] G. SORGER, On the minimum rate of impatience for complicated optimal growth paths, *Journal of Economic Theory* **56** (1992), pp. 160 –179.

[63] N. L. STOKEY AND R. E. LUCAS, JR., (WITH E. C. PRESCOTT), *Recursive Methods in Economic Dynamics*, Harvard University Press, Cambridge MA, 1989.

[64] P. A. STREUFERT, Stationary recursive utility and dynamic programming under the assumption of biconvergence, *Review of Economic Studies* **57** (1990), pp. 79 - 97.

[65] P. A. STREUFERT, An abstract topological approach to dynamic programming, *Journal of Mathematical Economics* **21** (1992), pp. 59 - 88.

162

[66] H. UZAWA, Time preference, the consumption function, and optimum asset holdings, in *Value, Capital and Growth: Papers in Honour of Sir John Hicks* (J. N. Wolfe, Ed.), pp. 485 - 504, Edinburgh University Press, Edinburgh, 1968.

Printing: Druckhaus Beltz, Hemsbach
Binding: Buchbinderei Schäffer, Grünstadt

Lecture Notes in Economics
and Mathematical Systems

For information about Vols. 1–210
please contact your bookseller or Springer-Verlag

Vol. 359: E. de Jong, Exchange Rate Determination and Optimal Economic Policy Under Various Exchange Rate Regimes. VII, 270 pages. 1991.

Vol. 360: P. Stalder, Regime Translations, Spillovers and Buffer Stocks. VI, 193 pages . 1991.

Vol. 361: C. F. Daganzo, Logistics Systems Analysis. X, 321 pages. 1991.

Vol. 362: F. Gehrels, Essays In Macroeconomics of an Open Economy. VII, 183 pages. 1991.

Vol. 363: C. Puppe, Distorted Probabilities and Choice under Risk. VIII, 100 pages . 1991

Vol. 364: B. Horvath, Are Policy Variables Exogenous? XII, 162 pages. 1991.

Vol. 365: G. A. Heuer, U. Leopold-Wildburger. Balanced Silverman Games on General Discrete Sets. V, 140 pages. 1991.

Vol. 366: J. Gruber (Ed.), Econometric Decision Models. Proceedings, 1989. VIII, 636 pages. 1991.

Vol. 367: M. Grauer, D. B. Pressmar (Eds.), Parallel Computing and Mathematical Optimization. Proceedings. V, 208 pages. 1991.

Vol. 368: M. Fedrizzi, J. Kacprzyk, M. Roubens (Eds.), Interactive Fuzzy Optimization. VII, 216 pages. 1991.

Vol. 369: R. Koblo, The Visible Hand. VIII, 131 pages.1991.

Vol. 370: M. J. Beckmann, M. N. Gopalan, R. Subramanian (Eds.), Stochastic Processes and their Applications. Proceedings, 1990. XLI, 292 pages. 1991.

Vol. 371: A. Schmutzler, Flexibility and Adjustment to Information in Sequential Decision Problems. VIII, 198 pages. 1991.

Vol. 372: J. Esteban, The Social Viability of Money. X, 202 pages. 1991.

Vol. 373: A. Billot, Economic Theory of Fuzzy Equilibria. XIII, 164 pages. 1992.

Vol. 374: G. Pflug, U. Dieter (Eds.), Simulation and Optimization. Proceedings, 1990. X, 162 pages. 1992.

Vol. 375: S.-J. Chen, Ch.-L. Hwang, Fuzzy Multiple Attribute Decision Making. XII, 536 pages. 1992.

Vol. 376: K.-H. Jöckel, G. Rothe, W. Sendler (Eds.), Bootstrapping and Related Techniques. Proceedings, 1990. VIII, 247 pages. 1992.

Vol. 377: A. Villar, Operator Theorems with Applications to Distributive Problems and Equilibrium Models. XVI, 160 pages. 1992.

Vol. 378: W. Krabs, J. Zowe (Eds.), Modern Methods of Optimization. Proceedings, 1990. VIII, 348 pages. 1992.

Vol. 379: K. Marti (Ed.), Stochastic Optimization. Proceedings, 1990. VII, 182 pages. 1992.

Vol. 380: J. Odelstad, Invariance and Structural Dependence. XII, 245 pages. 1992.

Vol. 381: C. Giannini, Topics in Structural VAR Econometrics. XI, 131 pages. 1992.

Vol. 382: W. Oettli, D. Pallaschke (Eds.), Advances in Optimization. Proceedings, 1991. X, 527 pages. 1992.

Vol. 383: J. Vartiainen, Capital Accumulation in a Corporatist Economy. VII, 177 pages. 1992.

Vol. 384: A. Martina, Lectures on the Economic Theory of Taxation. XII, 313 pages. 1992.

Vol. 385: J. Gardeazabal, M. Regúlez, The Monetary Model of Exchange Rates and Cointegration. X, 194 pages. 1992.

Vol. 386: M. Desrochers, J.-M. Rousseau (Eds.), Computer-Aided Transit Scheduling. Proceedings, 1990. XIII, 432 pages. 1992.

Vol. 387: W. Gaertner, M. Klemisch-Ahlert, Social Choice and Bargaining Perspectives on Distributive Justice. VIII, 131 pages. 1992.

Vol. 388: D. Bartmann, M. J. Beckmann, Inventory Control. XV, 252 pages. 1992.

Vol. 389: B. Dutta, D. Mookherjee, T. Parthasarathy, T. Raghavan, D. Ray, S. Tijs (Eds.), Game Theory and Economic Applications. Proceedings, 1990. ??, ?? pages. 1992.

Vol. 390: G. Sorger, Minimum Impatience Theorem for Recursive Economic Models. X, 162 pages. 1992.

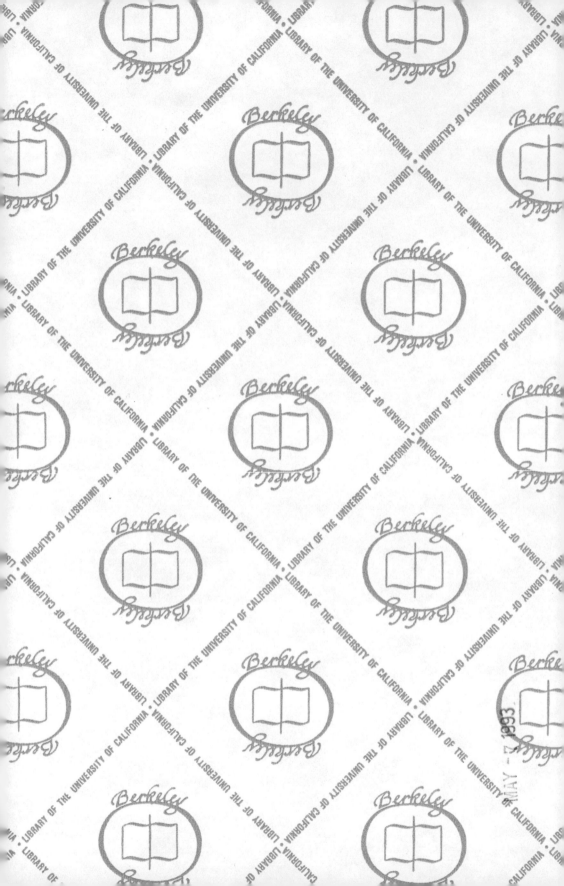